MIGRATION: THE ASIAN EXPERIENCE

Also by Judith M. Brown

GANDHI: Prisoner of Hope

GANDHI AND CIVIL DISOBEDIENCE: The Mahatma in Indian Politics, 1923–34

GANDHI'S RISE TO POWER: Indian Politics, 1915–22

MEN AND GODS IN A CHANGING WORLD

MODERN INDIA: The Origins of Asian Democracy

Also by Rosemary Foot

THE WRONG WAR: American Policy and the Dimensions of the Korean Conflict, 1950–53

A SUBSTITUTE FOR VICTORY: The Politics of Peacemaking at the Korean Armistice Talks

Migration: The Asian Experience

Edited by

Judith M. Brown
Beit Professor of Commonwealth History
and Fellow of Balliol College
University of Oxford

and

Rosemary Foot
John Swire Senior Research Fellow
in the International Relations of East Asia
and Fellow of St Antony's College
University of Oxford

M
St. Martin's Press in association with ST ANTONY'S COLLEGE, OXFORD

First published in Great Britain 1994 by
THE MACMILLAN PRESS LTD
Houndmills, Basingstoke, Hampshire RG21 2XS
and London
Companies and representatives
throughout the world

This book is published in the *St Antony's/Macmillan Series*
General Editor: Alex Pravda

A catalogue record for this book is available from the British Library.

ISBN 0–333–60100–9

Printed in Great Britain by
Antony Rowe Ltd
Chippenham, Wiltshire

First published in the United States of America 1994 by
Scholarly and Reference Division,
ST. MARTIN'S PRESS, INC.,
175 Fifth Avenue,
New York, N.Y. 10010

ISBN 0–312–09723–9

Library of Congress Cataloging-in-Publication Data
Migration, the Asian experience / edited by Judith M. Brown and Rosemary Foot.
p. cm.
Includes index.
ISBN 0–312–09723–9
1. Asia—Emigration and immigration. I. Brown, Judith M. (Judith Margaret), 1944– . II. Foot, Rosemary, 1948–
JV8490.M54 1994
304.8' 095—dc20 93–47025
CIP

Contents

List of Maps

Acknowledgements

We would like to express our grateful thanks to the Asian Studies Centre of St Antony's College for providing the opportunity to hold the seminar series on which this collective volume is based; to the School of Geography at Oxford for invaluable assistance with the preparation of the maps; and to the Inter-faculty Committee for South Asian Studies, Queen Elizabeth House; and the John D. and Catherine T. MacArthur Foundation for financial support. We would also like to thank the editors of the journal *Migration* for giving permission to publish a revised version of Dr Werbner's article which first appeared in August 1990; and Chloë Dobree and Marga Lyall whose administrative and typing skills greatly aided us in the final preparation of the manuscript.

JMB
RF
Oxford, October 1993

Notes on the Contributors

Sarah Ansari is Lecturer in Modern Middle Eastern and World History at Royal Holloway College, University of London. She is the author of *Sufi Saints and State Power: The Pirs of Sind, 1843–1947*. Her research interest in issues of migration began when she was awarded a British Academy Post-doctoral Fellowship in 1988. Since then she has worked on partition-related migration between India and Pakistan in general and to the Pakistani province of Sind in particular.

Marcus Banks gained his doctorate in social anthropology from the University of Cambridge and went on to train at the National Film and Television School, Beaconsfield. His original research was on the Jains of Gujarat and England and has formed the basis for a monograph *Organizing Jainism in India and England* (Oxford, 1992) and a series of articles. He currently teaches social anthropology at the University of Oxford and is conducting research on systems of visual representation.

Judith M. Brown is Beit Professor of Commonwealth History in the University of Oxford and Fellow of Balliol College and is closely connected with St Antony's College through the Asian and Indian Studies Centre. She has written widely on modern India; her most recent books are a biography of Gandhi and a new edition of *Modern India: The Origins of an Asian Democracy.*

Graham E. Clarke is a Fellow and Coordinator of the Anthropology and Development Programme at Queen Elizabeth House, Oxford His work includes anthropological research, as well as archival and statistical analysis. He has carried out a number of periods of fieldwork in the Himalayas and Central Asia since 1969, and research and advisory work for the Commission of the European Communities in the People's Republic of China, especially Tibet and Qinghai, at various times between 1986 and 1992.

Roger Daniels is Professor of History at the University of Cincinnati and vice-president and president-elect of the Immigration History Society (USA). His most recent books are: *Prisoners Without Trial: Japanese Americans and World War II* (1993); *Coming to America: Immigration and*

Ethnicity in American Life (1990); *History of Indian Immigration to the United States: An Interpretive Essay* (1989); and *Asian America: Chinese and Japanese in the United States since 1850* (1988). He has been a visiting professor at the universities of Hamburg, Innsbruck, Calgary and Toronto, an editor of three scholarly series, a consultant to governmental bodies, and has helped to produce television films.

Rosemary Foot is the John Swire Senior Research Fellow in the International Relations of East Asia and a Fellow of St Antony's College, Oxford. Her most recent book is *A Substitute for Victory: The Politics of Peacemaking at the Korean Armistice Talks* (lthaca, 1990) and she is currently working on a study of American relations with China since 1949.

John R. Hinnells, formerly Professor of Comparative Religion at Manchester University, is now Head of the Study of Religions Department, School of Oriental and African Studies, London University. His writings cover four main areas: Zoroastrian influence on the Judaeo-Christian tradition; Roman Mithraism (*Mithraic Studies*, 2 vols, Manchester, 1975 and *Studies on Mithraism*, Rome, 1984, both edited works); Zoroastrianism and the Parsis in India and overseas (*Persian Mythology*, London, 1985) and editorial work in the broad field of the study of religions including the *Penguin Dictionary of Religions* (1983); *Handbook of Living Religions* (London, 1984); *Who's Who of World Religions* (London, 1991).

Linda Hitchcox is a social anthropologist presently directing a development programme in Vietnam for a Scandinavian consortium of aid agencies. She was a senior member of St Antony's College from 1988 to 1992 engaged in research on migration and resettlement issues in Vietnam. In 1990, she published *Vietnamese Refugees in Southeast Asian Camps* (London).

Ceri Peach is Professor of Social Geography at Oxford University and a Fellow of St Catherine's College. He has held Visiting Fellowships at ANU, Berkeley and Yale. His research interests are in patterns of migration, settlement and segregation. His first book was *West Indian Migration to Britain: A Social Geography* (Oxford, 1968). He edited, together with Colin Clarke and Steven Vertovec, *South Asians Overseas: Migration and Ethnicity* (Cambridge, 1990). He is currently working on the growth of the Muslim population in Europe.

Yoko Sellek is a lecturer in the School of East Asian Studies at the University of Sheffield. She completed both her BA and MA studies at Sophia University in Tokyo, Japan, where she specialised in International Relations. Her current interests focus on the relationship between national integration and migration chains in Asia. She has recently conducted field work in Japan to investigate the structure of the networks of Korean permanent residents and Japanese day-labourers that are found in areas where illegal migrant workers reside.

Pnina Werbner is Senior Lecturer in Social Anthropology at the Department of Sociology and Social Anthropology, Keele University. She has published extensively on South Asian Muslims in Britain and in Pakistan, including *The Migration Process: Capital, Gifts and Offerings among British Pakistanis* (Oxford, 1990), *Black and Ethnic Leaderships in Britain* (London, 1991), co-edited with Muhammad Anwar, and *Economy and Culture in Pakistan* (London, 1991), co-edited with Hastings Donnan. Her current research is on Sufism as a transnational religious movement.

Edgar Wickberg is Professor Emeritus of Modern Chinese History at the University of British Columbia. His family background includes migration from Sweden to the United States and thence to Canada, and much of his research has been about the history and sociology of the global diaspora of the Chinese. Among his major writings are *The Chinese in Philippine Life, 1850–1898* (1965) and *From China to Canada: A History of the Chinese Communities in Canada* (1982) (editor and principal author).

Introduction: Migration – The Asian Experience

Judith M. Brown and Rosemary Foot

Human migrations are a significant force in historical change. Large-scale migration, of Asians, Africans and Europeans, was central in fashioning the world of the nineteenth century, its world-wide empires and the establishment of global economic and political dominance by people of European origin. Twentieth-century migrations have been just as important, particularly in the creation of a post-World War and post-imperial world order. Asian peoples have been among the most central in these processes. By the last quarter of this century about nine million South Asians (Indians, Pakistanis and Bangladeshis) and over 30 million Chinese are to be found living outside their country of origin as a result of migration or birth within a migrant community. There were also significant numbers of 'internal migrants', displaced within their home regions; such as the millions traversing the Indo-Pakistan borders after the partition of the subcontinent in 1947, and the Chinese who have moved into the predominantly Tibetan western reaches of the Chinese state.

Asian migration is, therefore, not only an important aspect of Asian history and the life experiences of people from that continent, but is also part of the creative contribution of Asia and its peoples to the contemporary world. Recognition of this is a valuable corrective to any over-simplified categorization of Asia as 'Third World' or 'dependent' and its peoples as passive or victimized. Certainly many migrants were victims – of natural and political disasters. But others who migrated were making conscious, positive decisions, and adopting innovative socio-economic strategies for themselves and their families. It was in recognition and exploration of the richness and diversity of the Asian experience of migration and its significance that we organized a seminar series in Oxford in 1991. That series, supplemented by several papers specially commissioned to provide geographical spread and important comparative material, form this volume.

The collection provides a picture of the main broad streams of Asian migration, and their wide geographical spread, both in terms of migrants' origins and their destinations. Evidence comes from several of the countries of South and East Asia. It shows migrants moving within their own countries (Banks, Clarke, Hitchcox); abroad but still within Asia (Ansari, Banks, Hitchcox, Sellek and Wickberg); and overseas, particularly to Britain and North America (Daniels, Hinnells, Peach, Werbner and Wickberg). Some of the studies are large canvases, depicting diaspora of peoples – Hinnells on the Parsis, and Wickberg on overseas Chinese communities. Peach looks at the broad trends of South Asian migration since the mid-nineteenth century. Other contributors focus on a smaller scale, on particular groups or local experiences. Werbner looks at Pakistani immigrants who carved out an economic niche in the economy of Manchester in north-western England, while Ansari analyses the experience of Indian Muslim refugees in Pakistan after 1947, and Banks focuses on several small groups of Gujarati Jains who migrated to East Africa and then further afield. Clarke provides significant new evidence on Han migration westward in China over a long period. Two interesting sub-categories or sub-themes in the pattern of out-migration from the homeland are re-emigration and illegal migration. Peach and Banks both discuss circumstances in which South Asian migrants have been forced to move on and become 'twice-migrants', East Africa being the most egregious example. The plight of illegal migrants is also noted: Vietnamese refugees forced to return from Hong Kong (Hitchcox) and illegal foreign workers expelled from Japan (Sellek).

Anyone looking at large-scale migrations will ask the question 'Why?'. Often the causes of migration are discussed in existing literature in terms of push and pull factors; and Hinnells and Wickberg both draw on this framework. However, Banks in the opening to his paper warns against over-simplification and crude assumptions about causation. He notes that anthropologists like himself (but he could equally well have included other related disciplines) tend to deal with movements or events which have happened, and therefore can err with hindsight into assuming that such trends or happenings were inevitable. Further, they tend to think of migrants as sociological groups and therefore miss the intricate patterns of decision-making which precede migration. There will certainly be broad disposing factors encouraging migration as an option. But it is individuals and often families who take the decision to migrate. Hitchcox shows this familial decision-making process at work in Vietnam. The common pattern among South Asians of 'chain migration' likewise points to the importance of families and kin in the process of election for the migration strategy. This leads us to ask the reverse question: why do some people choose *not* to migrate in what are apparently similar circumstances? The evidence that quite small areas produce different patterns of choice about whether to stay or go, among very similar types of people (as Banks shows among Gujarati Jains) reinforces the need for such caution in discussing causes, and for more case-studies both of migrants and non-migrants who share a common home environment yet decide to manage their lives in different ways.

One major theme which emerges from several of these studies is that, whatever the immediate factors making migration seem a necessary step, or a positive and desirable option, many of those who migrate come from regions or groups where there is a tradition of movement, often stretching back centuries. The Parsis, Hinnells argues, have an ancient migration religion. From the time of Zoroaster they have been on the move, but particularly since they fled from persecution in Persia in the tenth century to India. The great Parsi diaspora of the last two centuries from South Asia therefore falls into an established community pattern of response to both danger and opportunity. The Jain migrants to East Africa and Britain (Banks) are likewise acting within an established pattern of behaviour preserved in communal memory which started nearly five centuries ago as groups of Jains moved from Rajasthan into Gujarat and Bombay.

Contemporary Vietnamese movement within and outside Vietnam is the latest aspect of many decades of political, civil and economic disruption in Vietnamese society which has forced people to move. They have acquired what Hitchcox calls 'the habit of movement'. Wickberg similarly describes the Chinese as 'a migrating people', despite a deep attachment to home.

Understanding why Asians have moved in such numbers and often so far demands consideration of a range of predisposing factors, both objective and subjective: what people were experiencing, how they understood and interpreted those experiences and how they perceived opportunities and possibilities away from home. Clearly economic and demographic factors were significant for many migrants: opportunities thought or known to be available abroad or elsewhere in the state as opposed to those at home which were seen as limited, stagnant or declining. Han migration within China, and South Asian movement overseas are obvious examples. It is fairly unusual for economic disaster or abject poverty to be the 'cause' of voluntary migration, compared with the domestic conditions which induced Indians to participate in the indentured labour scheme of the nineteenth century. Free migrants need initiative and resources to leave. But in some cases absolute as opposed to relative or perceived poverty and deprivation has been important. The great famine of 1899–1901 in Gujarat was a strong impetus to Jain migration (Banks), as it was to many Patidars, a peasant caste prominent in Kaira district of central Gujarat. Much

Vietnamese movement similarly occurred in the context of extreme economic dislocation, falling soil fertility and food shortages. (See Hitchcox's analysis of 'migrant generating regions'.) More often it is the apparent opportunities abroad which persuade people to leave home, for individual and familial advancement in a new area, often combined with the hope of sending back remittances to enrich the extended family and possibly to provide for an individual's comfortable old age in his or her home village. As Sellek points out, Japan's wage levels are far above those of neighbouring countries. (One hour of wage labour in Japan, for example, is many times that of neighbouring Asian countries. Other Asians often perceive the streets of Japan to be paved with gold.)

So there are accounts here of streams of migrants who leave home in search of a variety of socio-economic El Dorados. Some have been students or those who are already well educated and qualified professionals who leave in search of improved professional opportunities and the chance of even better education for their children. The Parsis around the world, the later twentieth-century Asian Indian and Chinese communities in North America are such (Hinnells, Daniels and Wickberg). They can also be entrepreneurs of various kinds, ranging from big businessmen and large-scale traders to the ethnic restaurateur or the family who run a corner shop in Britain. Werbner's account of Pakistani entrepreneurial initiative is an example of this type; so are Banks' Jains, and Wickberg's Chinese. Other Asians have moved with only their labour to sell in a new environment. This had been the case of Indian indentured labourers in the nineteenth century as Peach notes. Sellek's chapter on migrant workers in Japan shows how both skilled and unskilled flock to an Asian capitalist metropolis in search of work, providing labour and services the local society cannot or will not generate itself. Clarke's chapter demonstrates that economic reform policies in the post-Mao era have encouraged Han migration into Tibet where these new migrants work in construction, petty trade and semi-skilled service industries.

Many of these studies (like Clarke's) indicate that political factors of various kinds are also significant in 'persuading' people to migrate and in controlling the type and numbers of migrants at the receiving end of the process. Civil strife was of primary importance in at least three of our cases. The Indian Muslim migrants into the new state of Pakistan at or soon after 1947 were people who felt they had no choice but to leave home when religious communalism and violence racked northern India and the British devolved power to a predominantly Muslim Pakistan, and a secular (but mainly Hindu) India. For them the alternative to migration seemed to be slaughter and dispossession, or at the very least discrimination in a profoundly hostile environment. Many Chinese left the mainland once the Communists assumed power in 1949. Civil war and its aftermath was also central to the developing patterns of Vietnamese population movement later in the century.

Actual persecution, harassment and discrimination have also pushed some migrants. Hinnells shows how twentieth-century Parsis fled from Iran after the fall of the Shah, and have left Pakistan for fear of 'Islamization' of the polity. Often pressures have occurred in situations where Asians had established themselves in the entrepreneurial sectors of a local economy, the Chinese in Thailand, for example; or in the professions and administration, often under a British colonial regime which welcomed their skills and loyalty. For this last group, when independence came local people resented these migrant colonial 'collaborators' and harassed or evicted them. Peach notes this happening in Sri Lanka, Burma and East Africa. In Surinam, by contrast, South Asians were lowly people, indentured labourers and their descendants; but they, too, fled for political reasons – the prospect of Creole domination in the 1970s.

In other contexts governments have found migration a useful tool in the building of nation-states or even ideologically based international communities and have directed the movement of peoples to serve the ends of security, economic development and proletarian internationalism. Hitchcox and Clarke on Vietnam and China both underline the importance of such government policies. Before 1987 the Vietnamese government attempted to control all internal and external population movement, directing people for the state's purposes, particularly to exploit new economic zones and to relieve overcrowded regions. The political benefits of mixing a society which had been ideologically divided and of moving loyal citizens to border regions were also official goals. The success of such officially prompted migration is, however, questionable. In China the PRC government has similarly used population movement to extend both control and economic change in the north-west and in Tibet. In both China and Vietnam, however, governments have in the last decade relaxed their controls on movement, and therefore significant voluntary migration is now occurring, often for broadly economic reasons.

However, totalitarian regimes are not the only ones which seek to control migration, as Japan, Australia, North America and Britain have demonstrated. Virtually all modern nation-states exercise controls on immigration as they mark out the boundaries of national identity, watch that an influx of outsiders does not overstretch available resources, complements rather than competes with locals and their skills, and attempt to prevent other governments exercising leverage in domestic circumstances. Official controls on immigration which have markedly altered the flows of Asian migration receive some attention here. Peach notes the severe tightening of British nationality and immigration law from 1962 onwards, which stemmed the stream of South Asian migrants into Britain. Daniels traces the transformation of United States immigration laws in 1965 and the opportunity this gave for a striking increase in the numbers of Asian Indians in America with skills and qualifications the government felt to be desirable. Sellek in contrast to Daniels notes the tightening of procedures in Japan as a result of the social pressures being generated by Asian migrant workers.

Many have made a long-term success of migration, but it is worth noting those who returned to their country of origin, often as a result of political pressure. Migrants often hold on to a 'myth of return' for a generation at least, though very few actually do leave their newly acquired homes and identities, particularly if they have raised their families in their new homes and invested both social and economic capital there. Banks, however, does describe Gujaratis expelled from East Africa who chose to return to India rather than move on to the UK or the USA. Back 'home' in Jamnagar, Gujarat, these returned migrants live in a separate residential area, characterized by large ornate homes and new Jain temples, deploying their African capital in entrepreneurial agriculture or speculation in the city's futures markets. They are somewhat resented locally as flashy upstarts. (Since East African Asians in Britain are similarly regarded by other Asians, it is perhaps a reflection of a distinctive African-Asian style rather than a function simply of being a returnee.) By contrast in Vietnam, when 'failed migrants' with no such pretensions return home, they appear to have been received locally with considerable tolerance, warmth and understanding of their predicament.

Most Asians who migrate do so permanently and are often joined by their extended families and neighbours when immigration law permits. Their migration is therefore presumably to some extent 'successful'. But their experience of uprooting and movement and re-rooting must be recognized as a *process*. This process takes time, develops and unfolds over several generations, as both in-comers and the host society respond to each other's propinquity. In many cases those who have contributed to these studies of Asian migration as well as those who read them are themselves therefore involved in this process. The consequences are deep and long-lasting, both for those who move and those who receive new neighbours. The receiving or 'host' societies are often profoundly influenced by the inflow of new residents and citizens. Migrant labour and migrant business initiatives, as well as immigrant participation in the professions, are often economically significant. (See Werbner, Daniels and Wickberg, for example.) The presence of new religions and cultural groups provides the challenge of a potentially multicultural society, which may be interpreted as a threat or an invitation to cultural tolerance and creativity, an issue Hinnells notes. Almost inevitably the presence of immigrants, short and long-term, generates new types of tension within the receiving society and between the governments of the countries which export and import migrants – tensions which demand social and political resolution. From the point of view of those who migrate there is not just the question of 'making good', economically, socially or politically, in a new home, and deciding whether migration is temporary or a permanent uprooting. There is also the more complex and subtle business of carrying the valued baggage of culture, tradition and often religion, and then making choices about the perpetuation or modification of such visions of life and patterns of value and appropriate behaviour in the new environment. Inevitably, too, long-range movement raises questions to do with social structure and relations between families and kin groups, between the sexes, and between generations. Such issues are central to the discussion by Hinnells and Wickberg.

Our case-studies do not attempt a comprehensive survey of the 'consequences' of migration. But they do raise and illuminate many questions about significant aspects of the migration experience. Most of them note the economic consequences of migration for the groups studied, though only Werbner takes this as her major theme. She shows how Pakistanis have come to dominate an enclave in the economy of Manchester and north-western England – a success story in an environment of industrial decline. British Pakistanis have now virtually taken over the cheaper end of the clothing trade and Werbner examines the networks of people, skills, capital, credit and knowledge which have brought this about. Another case of migrants coming to dominate a particular economic niche is mentioned by Daniels. Gujaratis now control a large part of the American motel and petrol station trade. Again, the significance of personal connections, transmission of knowledge, low capital start-up costs and labour intensive enterprise are themes which echo many of the points made by Werbner. Sellek's study of foreign labour in Japan also traces economic adaptation to a new environment: and the niche or enclave theme recurs with evidence of migrant female domination of the euphemistically named 'entertainment industry'. Ansari notes how Indian Punjabi migrants to Pakistani Punjab were predominantly agriculturalists who settled easily and naturally into a similar socio-economic status and life across the border. Their comparatively tranquil experience of transition was in stark contrast to that of Indian Muslims who went to Sind province only to be faced with a dire shortage of resources (land, houses, jobs), and very considerable local hostility; though ultimately many of them eventually took a dominant role in the developing economic life of Sind's towns, particularly Karachi.

Some Asian migrants have also been outstandingly successful in professional life and in achieving higher levels of education for their children than the average in the host society. The Parsis carry on a tradition of high levels of education for men and women wherever they go, which helps to account for their success in the professions, as in North America. But the Asian Indian and Chinese communities as a whole in the United States also show this investment in education and often success in professional life, as Daniels and Wickberg illustrate. The evidence gathered here suggests that there is no single pattern of migrant economic adaptation: rather the pattern in each context depends on the interplay of local opportunities and resources and the skills which migrants bring or are prepared to learn on arrival. Often being an 'outsider' may have its advantages for engaging in particular types of economic enterprise, even where immigrants find it difficult to break into the full range of economic activity in the host society.

However, 'making good' at a material level is only part of the ongoing business of migration. Each individual and family group in each generation has to face the question of the value of the original migrant culture and its contemporary relevance. Asians abroad must decide whether to keep all or part of their culture, belief systems, worship patterns and traditions of domestic interaction: they have to respond to the pressures and opportunities of the society and culture in which they find themselves, particularly as their children are educated in the host society's educational systems. Hinnells sees this as a particularly urgent concern for Parsis, whose home bases in Iran and South Asia are either demographically or culturally at risk. Our cases provide no evidence for any theory of cultural assimilation. What seems clear is that most Asians who migrate continue to value their inherited and imported culture and its manifestations very greatly and take pains to ensure its survival and inheritance by the younger generation, though there is considerable willingness to accept selectively from other traditions and practices. The Parsis for example watch and evaluate carefully the likely response to them of the societies where they go and the implications for living as good Parsis in those societies. But their attitudes to their own culture and religion are not themselves uniform, though they are a comparatively small and culturally homogeneous community. Different groups of Parsis are influenced in their interpretation of their own tradition by the societies from which they come – Iran, Karachi or Bombay, for example. The overseas Chinese, too, as Wickberg underlines, often continue to feel the pull of China and still value Chinese culture. Many migrants from Asia strive to retain their culture by maintaining strong kinship ties, ethnic marriage networks, and often residence in ethnic enclaves. The Chinese have developed a supporting range of ethnic associations, a parallel to the temple and mosque associations developed among South Asians in Britain. (Though it is noteworthy from other studies of migration that these associations can also be the focus of conflict within migrant communities as different leaders and factions struggle for control of community resources and the right to speak for the migrant group to the outside society from a legitimate institutional base.) Clearly numbers and concentration are important factors in migrants' cultural responses to their new position. Wickberg points out some of the implications of these differentials for overseas Chinese who find themselves in very different types of situation abroad: as a majority in Hong Kong, as a large minority in Malaysia, and as small minorities in Britain and elsewhere.

A theme which emerges from these studies but needs more investigation is the effect of migration on familial relations. Integral to this is the self-perception and role of women and of the second and succeeding generation of migrants who grow up in the new environment. Ansari notes, for example, that Indian Muslim women coming into Sind tended to be socially freer, better educated and more involved in public life than their Sindhi counterparts: and as a result of their arrival the position of women in the province began to change. This must be quite a rare case of migrants actually changing patterns dominant in the host society. Hinnells and Daniels note the high education levels of women and of female employment among Parsis and Indians in North America. This is in part the result of favourable opportunities resulting from migration; but it also reflects migration by groups which already value and try to ensure education for their daughters, and view paid work outside the home as socially acceptable. There is apparently therefore no easy and simple equation between migration and the expansion of roles and opportunities for women. We need to know far more from a range of comparative migration situations about whether migration tends to extend the range of women's paid employment and educational opportunity. From opposing ends of the occupational and educational spectrum, Sellek's evidence of women in the Japanese sex industry and Daniels' note that Indian women in North America tend to cluster in health-related occupations suggest that women are still heavily in gender-typed occupations. The experience of younger people of both sexes is also crucial to the long-term experience of migration. Hinnells and Wickberg discuss directly the issue of whether a younger generation in a migrant community is likely to break away from tradition and inherited culture, and the tensions this generates within the individual and the family.

Clearly then, how migrants handle their own culture is a complex issue and the patterns vary and are still under negotiation, as it were, in places where migrants are establishing themselves. Urbanization, education, mass communications and the demands of citizenship all necessitate selective cultural adaptation to the new environment. But one significant element in this process of interaction is obviously the reactions of the host society to the migrants. 'Receivers' as much as 'incomers' have to respond and adapt to the presence of the new and often misunderstood in their midst. Chapters which take this as an important theme are Ansari's analysis of the different experiences of Indian Muslims moving into Sind and Punjab provinces of Pakistan and Sellek's examination of Japanese society's response to the migrant worker. Ansari's example is especially significant because at the outset the rationale of the state of Pakistan and the shared religion of Islam would suggest that there would be little tension in accepting fraternal refugees in the new Islamic homeland. But this was not so. Although Indian Punjabis who settled in Pakistani Punjab did so with comparative ease, in Sind acute tension arose and still persists between ethnic Sindhis and the incomers, the *muhajirs*. Ansari describes a build-up of cultural, political and economic hostility towards the *muhajirs* and her evidence shows how much the reception of immigrants depends on the weight of numbers involved, the type of incomer and the resources available to meet their needs; but of paramount importance is the way in which the situation is managed within the political system. In the Pakistani case she shows how the state has fallen short, not only in the management of tension and the distribution of scarce resources, but in failing to create a new Pakistani identity which could subsume loyalties of ethnic origin.

Other migrants' experiences show similarities. Where Asians were constrained to leave East Africa there were ethnic and cultural tensions, issues of job opportunities and control of resources, and a nation-state incapable of or unwilling to invest in a new inclusive national identity and consensus. The cultural strains between in-comers and locals, seen between Sindhis and *muhajirs* despite their common religion, are evident to an even greater extreme where there is deep and obvious cultural difference. Hinnells looks at Parsi experiences of discrimination, while Werbner has some fascinating and disquieting evidence of institutionalized discrimination against Asians in the British insurance industry. Sellek notes the extreme cultural hostility in Japan to foreign workers.

This volume raises important questions integral to the migration process and experience. It provides evidence on many of them and by implication it suggests where more case-studies are needed. Because migration is a 'whole life' experience, understanding it demands the skills of a range of scholars. Represented here are sociologists, anthropologists, historians, comparative theologians, geographers and those who study international relations. Consequently we offer it to an equally broad audience, hoping too that non-specialists may also find here food for thought as they contemplate themselves and their neighbours and the societies and polities they are in the process of creating. Much existing literature on migration tends to be theoretical or is quite narrow in its focus on particular groups of migrants or particular situations in which migrants find themselves. By looking broadly at types and aspects of Asian migration over the last hundred years and more we offer a wider canvas, material for comparison, and a sense that what is happening now is a phenomenon not only of immense contemporary significance but which stands in a long historical continuum. It is also an examination of one way in which Asia and its peoples have made a vital contribution to the changing demographic, economic, cultural and political patterns of the contemporary world.

1 The Chinese as Overseas Migrants

Edgar Wickberg

MIGRATIONS

There are an estimated 30 million ethnic Chinese living outside China.[1] In the 1990s such expressions as 'The Chinese Diaspora' or 'Global Chinese' are widely heard, and few countries do not have at least a small ethnic Chinese population. The global spread and often entrepreneurial character of overseas Chinese populations have attracted the attention of governments and the media, while the world-wide Chinese-language media chronicle their doings and keep them in touch with one another.

The current condition of Chinese migrants is linked to and grows out of a long tradition of migration, both in China and abroad. The Chinese are both a migrating people and one with a deep attachment to home place, a seemingly contradictory pair of statements which provides insight into the historical attitudes and behaviour of migratory Chinese. This chapter examines some of the repercussions of this interaction of movement and desire for rootedness.

Migration shaped much of China's history and its dynastic cycles can also be seen as migration cycles. The early stage of each dynasty was marked by geopolitical expansion accompanied by migration. During periods of dynastic effectiveness and political control dynastic governments often sponsored Chinese migration and settlement on the frontiers. When a dynasty waned and frontiers contracted, Chinese migrants withdrew to safer territory. A new dynasty brought new territorial conquest and new migration. Yet, in seeming paradox, Chinese dynasties attempted to fix the residence of those they did not encourage as migrants. The ease of collecting taxes was a reason for that; but so was adherence to officially-supported Confucian norms that stressed family continuity, filial piety and respect for tradition. Although Chinese might welcome migration, having migrated, they tended to maintain their old 'home-place' registration for generations thereafter.[2]

Intersecting with such cyclical patterns of population movement were sustained changes over the last 900 years of Chinese history. Beginning about 1100 AD commercial development and urbanization began to bring about serious changes in China: 'Late Imperial China' (c. 1100–1900) was the result of these initial changes, followed in the sixteenth and seventeenth centuries, by sustained and rapid population growth. Cutting across these developments was the increasing involvement of China in the world economy. Sustained external migration by the sixteenth to eighteenth centuries, and especially since the mid-nineteenth century, is a result of the interaction of all these forces.

Thus, for the Chinese, overseas migration in the past 150 years is more than the consequence of involvement in the capitalist world system, or of colonialism. It is also the result of long-standing domestic changes in China itself, leading to both economic opportunity and demographic pressures that encouraged migration.

Thus, in conventional if somewhat unsatisfactory terms,[3] 'push' and 'pull' factors activated Chinese overseas migrations. The immediate 'push' stimuli of the nineteenth century were the dislocations caused by the Opium Wars, the Taiping Rebellion and other massive sociopolitical upheavals, the intrusion of the world economic system into coastal China and the growing inability of Chinese governments to cope with these and other developments. By the twentieth century civil war, Japanese invasion and revolutionary change became added pressures towards emigration.

On the 'pull' side there was the attraction to impoverished nineteenth-century Chinese peasants of labour demands in colonial South-East Asia and the Caribbean, and opportunities to build railroads in places like North and South America. The infamous coolie trade flourished between the 1840s and the 1870s, principally to serve the labour needs of mines and plantations. Alongside that there was an indentured labour system by which immigrants had first to repay their steamer-passage tickets before being free to go out on their own in the new land. There was also 'chain migration' of individual family members, the (usually male) pioneer member either sending for a teenage son or nephew, or taking one back with him after one of his own periodic visits to China.[4]

Besides mines, plantations and railroads, and ubiquitous opportunities for small commerce and artisanry, there was the lure of gold in North America and Australia. To these various attractions were added twentieth-century opportunities for modern education for those Chinese who were positioned to take advantage of them, and also the attractions of overseas asylum for many Chinese political exiles spawned by decades of war and revolution at home.

It needs to be stressed, however, that the resultant patterns of migration and adaptation were not only the workings-out of the immediate 'push-pull' forces of the nineteenth and twentieth centuries but the long tradition of migration as a family strategy for meeting adversity and seeking opportunity. Nowhere was this tradition so well established as in the two south-east coastal provinces of Fujian and Guangdong. From at least the sixteenth century a small number of counties (perhaps ten) in coastal Fujian had established regular patterns of overseas migration, family separation, remittance receiving and dependence. By the end of the eighteenth century a somewhat larger number of counties in Guangdong province were doing the same. In both provinces there came to be 'emigrant communities'[5] made up essentially of the wives, small children and elderly dependents of men and teenage boys working overseas – males who would probably make periodic visits and might eventually return to the village to retire.

Several other Chinese provinces had lesser traditions and experience of migration. The south-west provinces regularly sent migrants overland to parts of mainland South-East Asia. For the eastern coastal provinces of Jiangsu and Zhejiang, as for Fujian and Guangdong, overseas emigration was related to overseas trade.[6] Thus, people from Ningbo and the Shanghai area, long active in trade with Japan, were also among the many regional types of Chinese migrants in Japan. By the late nineteenth century Zhejiang people also found their way to continental Europe, following certain specific lines of export goods sales. Further north, in Shandong and the provinces of the North China Plain, it was China's own undeveloped north-east that drew attention in the early 1900s and these provinces did not become significant senders of emigrants to overseas destinations.

There have been three main 'waves' of migration since 1850: 1850–1900, 1900–49 and 1965 to the present.[7] In the late nineteenth century the pushes and pulls noted above produced the coolie trade, the 'credit ticket' (steerage passage) system on steamships after 1870, and a continuation of the earlier 'chain migration' method. Until 1850 nearly all Chinese overseas migration had been to South-East Asia on a chain migration basis. Much of the post-1850 migration went to the same places in South-East Asia but served to replenish communities already there. Those communities were made up of the residue of earlier immigration tides and their adaptation to South-East Asian environments had made their creolized societies something quite different from the newcomers of 1850 and after.[8]

Between 1850 and 1900 Chinese migrants also went to other areas for the first time: Australia, New Zealand and the Western Hemisphere. It was also in this period that subethnic variation among Chinese migrants became salient. Although only a few counties in the two provinces of Fujian and Guangdong regularly sent emigrants to South-East Asia, those two provinces are extraordinarily diverse in ethno-linguistic terms. By 1900 five major ethno-linguistic subgroups had emerged among South-East Asia's ethnic Chinese: the Hokkiens of Fujian and four subgroups from Guangdong. These four were the Cantonese, whose home localities were near Canton (Guangzhou) City and the Pearl River Delta in central Guangdong, the Chaozhou (Teochiu) people of eastern Guangdong, centred on Chaozhou, Shantou (Swatow) and environs; the Hakka people from widely dispersed areas of eastern and central Guangdong; and the Hainanese from Hainan Island just off the south coast. These ethno-linguistic subgroups established in each case their own preferred migration destinations and occupational niches abroad.

While South-East Asia became an arena of activity for all five subgroups, the new destinations of the late nineteenth century – Australia, New Zealand and the Western Hemisphere – became overwhelmingly Cantonese targets. It was Cantonese who sought gold and built railroads. But the Cantonese of the new destinations also engaged in shopkeeping and miscellaneous trades just as the South-East Asian Chinese did.

Chinese migrants of this era, whether to South East Asia or North America, were typically single males who rarely had their families with them. The question whether they and their successors in the early twentieth century were voluntary or involuntary sojourners has been controversial[9] but it may be enough here to say that sojourning seems to have been the outcome of the combined traditions and expectations of kin in China and the unwelcome reception Chinese often received abroad.

By the second wave (1900–49) the coolie trade had long since been terminated but the other two types of passage continued. One radically different development was the growing migration of women and children and the reconstitution of families abroad. The trend began around 1900 and became marked during the 1930s. The usual explanations of this tend to focus on Chinese government policies and on political instability in China, especially after the Japanese invasion of the north-east (Manchuria) in 1931. Whatever the reasons, family life overseas became somewhat more common than before,[10] though the sojourning single man or the father-son-nephew family fragment continued to be more common in most places.

Three other new types of immigrants appear: the overseas student, the Chinese schoolteacher-cum-journalist and the political exile. After 1900 Chinese governments, in quest of various forms of modernization and the skilled individuals appropriate thereto, sent substantial numbers of students to Japan and the West for higher education. Many families in China also saw overseas education as a new kind of migrant opportunity and family strategy. In the period 1900–49 most of these returned eventually to China, although political instability there drove some to shuttle between China and certain Western countries.

The Chinese schoolteacher-cum-journalist was the result of the appearance of Chinese schools in overseas Chinese societies during this period. Given the presence of more children in overseas societies, changing attitudes toward modern education in China and overseas, and the nationalist and other political aims of early twentieth century Chinese governments, overseas Chinese schools, frequently with Chinese government support and partial control, appeared in growing numbers, especially in the 1930s. The schoolteacher, usually from a different regional background and better educated than most overseas Chinese, was an outsider when he arrived. But he had the respect Chinese give to the educated and he made himself useful both as teacher and as mainstay of the Chinese newspaper of his locality. The teacher-journalist became the resident 'intellectual' of an otherwise business community. He taught Chinese language and cultural values to the immigrants' children, and wrote essays on those subjects and on China's politics, literature and history for newspaper readers. Since both schools and newspapers concentrated on China and on local Chinatown affairs, rather than the affairs of the country of residence, the teacher-journalist could be an important figure in the maintenance of Chinese values and a China orientation.

The political exile, like the overseas student, was a temporary member of the overseas society. He was usually well educated and he devoted his time to promoting political involvement with China. Thus he became another component of the small but respected (and sometimes influential) 'intellectual' stratum of the overseas society.

World War II in Asia and the subsequent political struggle in China resulted in a new wave of refugees – economic as well as political. By 1949 there were probably something like 10 million ethnic Chinese living outside China.[11] South-East Asia, as before, was the major settlement area, with North America (the United States and Canada) as second. There were smaller numbers of Chinese in Latin America, Western Europe and many scattered locales. Eighty-five per cent or more of the ethnic Chinese lived in South-East Asia. Though the term 'Gold Mountain' had been applied wishfully by emigrants to North America and Australia, there were no rich Chinese in either place. The Chinese emigrant's dream of rags to riches, when realized at all, was acted out only in South-East Asia.

In the 20 years following World War II (1945–65) some major changes in immigration occurred. The outflow of Chinese from the mainland to Hong Kong did not, as it had in the previous century, result in a further outflow to South-East Asia. The countries of South-East Asia, most of them newly independent, generally cut off Chinese immigration or reduced it to a trickle. Partly this was a matter of nationalist resistance to further Chinese economic power in this part of Asia. It also, however, reflected South-East Asian concerns after 1949 about the appearance of an assertive and possibly expansionist Communist China next door. The result was that the South-East Asian Chinese societies became increasingly populations of mostly local-born Chinese with less intimate ties to China and increasingly numerous links to South-East Asia.[12]

Although some Hokkiens managed to reach Manila from Fujian via Hong Kong,[13] direct emigration from the mainland virtually ceased and the majority of emigrants out of Hong Kong were Cantonese. Their destinations were primarily in North America, where both the United States and Canada had changed their immigration policies to allow Chinese families to be reunited, and Britain, which offered business opportunities to well-established Hong Kong Cantonese. As refugee Cantonese poured into Hong Kong from neighbouring Guangdong province in 1949 and again during China's Great Leap Forward in the years around 1960, some Hong Kong Cantonese found it advantageous to lease their farmland in the rural part of the Colony to the newcomers, going off themselves to engage in the Chinese restaurant business in London.[14] But this migration was not enough to create a 'new wave' as yet.

Another new development of these years was the emergence of a second source of Chinese overseas migration: Taiwan. The end of the Chinese Civil War in 1949 left Taiwan as a *de facto* independent entity and a competing 'China'. Taiwan's Cold War affiliation with the United States, economic improvements on the Island itself, and America's willingness to accept Taiwan students led by the 1960s to a growing body of Taiwanese youth who sought advanced training in American universities. Since the majority did not return, there developed over the next 30 years an important Taiwan (Hokkien and Mandarin speaking) component to the otherwise Cantonese population in the United States and an enrichment of Chinese America's intellectual stratum.

The third wave of Chinese overseas migration began about 1965. It was shaped by economic and political conditions in the sending areas, policies in the receiving areas, and new technological developments in transportation and communication. Political instability and danger in mainland China during the Cultural Revolution years encouraged migration where such was possible, as in the case of Canada's agreement with China for direct family-reunification migration. In the other two sending areas, Hong Kong and Taiwan, growing prosperity and indications of uncertain political futures caused newly middle-class Chinese to look about for safer, more stable places to protect their assets and to raise their children. Particularly critical were the Hong Kong riots of 1967, in which the spillover of the Cultural Revolution into the Colony caused many there to begin to consider migrating strategies and options.[15]

In the mid-1960s the United States and Canada revised their immigration policies to allow a much larger volume of Chinese immigration. Migration to those places was now no longer to be tied exclusively to family reunions: qualified individuals without relatives could migrate on their own. In subsequent years both countries increasingly wooed Chinese and other immigrants with skills and money to invest. At about the same time Australia abandoned its 'White Australia' policy, moving in the same direction as the North American countries.

These policies suited the very middle-class people who wished to leave Hong Kong and, to a lesser extent, Taiwan – people whose affluence had allowed them to acquire professional, technical and language skills that made it relatively easy for them to move into attractive positions in English-speaking countries. Moreover, the more affluent the families the larger were the number of children able to study abroad – even if the family did not immediately emigrate with them – especially in Australia, Canada and the United States. Added to these were the swelling numbers of mainland Chinese students in Western universities after 1980, some of whom became political or economic exiles, especially after the Tiananmen crackdown of 1989. By the 1980s the Anglo-Chinese agreement on Hong Kong's future urged forward the pace of international familial migration strategy arrangements in Hong Kong and put Taiwan's comparable population on notice that its own future was not certain.

The economic growth of Hong Kong and Taiwan, and the business expansion possible through enhanced transportation and communications technology also encouraged transcendence of the boundaries of Taiwan and Hong Kong and creation of global networks of kin and business, often at least partly synonymous. The middle-class Chinese family with several branches in key cities around the world, and the 'spaceman' entrepreneur who spent his time in transcontinental flights between the loci of his family members and those of his various business interests, became familiar figures.

As families became internationalized, with individual branches striking roots in key world cities but all maintaining touch with each other, the Chinese populations of those cities became more diverse in ethnolinguistic terms. In North America in particular, Hokkien, Hakka, Chaozhou and other non-Cantonese subethnic restaurants, associations and churches found places alongside older (and newer) Cantonese establishments. America's Chinese population increased to over one and a half million,[16] in a total US population of about 250 million. Canada, with roughly one-tenth the American population, had about half as many ethnic Chinese (700 000). In both countries Chinatowns became vibrant again and Chinese restaurants and other forms of cultural influence were found throughout the metropolitan areas of most cities.[17]

Things happened differently in South-East Asia. There was little new immigration but the Chinese already there took advantage of global economic opportunities, especially in Singapore. Instead of being a receiving area as before, South-East Asia now became to some extent a sending area for Chinese. As in Hong Kong and Taiwan there was now a significant group of Chinese with education and portable skills. In particular, the English-speaking middle-class Chinese of Singapore, Malaysia and the Philippines were positioned to take advantage of the attractions of the English-speaking receiving countries (Australia, New Zealand, Canada and the United States); they were encouraged by local policy and other irritants to do so. Singapore Chinese found government over-regulation confining; Malaysian Chinese were provoked by the race riot of 1969 and subsequent government policies aimed at redistributing access to opportunities; Philippine Chinese, finally given Philippine citizenship, found it did not protect them from kidnapping and extortion. The ensuing migration added to the subethnic complexity of Chinese societies where migrants settled. The original South-East Asian ethno-linguistic subdivisions were no longer conclusive once one had migrated again. In Vancouver a Singapore Hokkien is not the same as a Manila Hokkien or a Taiwan Hokkien.

Despite the rapid growth of the Chinese population in North America over the past 25 years, well over 80 per cent of the world's 30 million ethnic Chinese outside China still live in South-East Asia. Most are local-born and consider themselves rooted in South-East Asia, but not without the option to move on and not without familial and emotional ties to home in China. North America's Chinese, now, as the result of massive immigration since the 1960s, mostly foreign-born, also consider themselves more or less rooted, but with the options and sentiments mentioned above and with considerable differences between newcomers and oldtimers on these and other issues.

ADAPTATIONS

Many things condition overseas Chinese adaptation to local environments. One of them is the timing, composition and size of Chinese settlement, especially in relation to the settlement of non-Chinese in that location. We might consider the following as demographic types: (1) Majority Chinese societies; (2) Large Minority Chinese societies; (3) Minority Chinese societies.

The 'national' level Majority Chinese societies would include Taiwan, Hong Kong, Macau and Singapore. Although Taiwan was politically part of China before 1895, it has been politically and culturally separated from mainland China for the last century, half of that time under Japanese rule, the other half under Chinese rule. With acceptance of some ambiguity and without prejudice to its present claims and future status politically, we can regard Taiwan as 'offshore' or 'overseas'. The population of Taiwan is in effect entirely Chinese. A Hokkien population made up of descendants of seventeenth and eighteenth century immigrants from the mainland makes up 85 per cent of a population of 20 million. Culturally and politically superimposed on that is a 15 per cent proportion made up of mid-twentieth century arrivals from the mainland, bringing with them a Mandarinized Chinese culture from the Lower Yangtze River region. Before their arrival Taiwan was subjected to several decades of Japanese cultural influence and since then American influence has been strong. This is, therefore, a Chinese society with a Chinese government but with a peculiar political and cultural experience.

Hong Kong, founded in the 1840s, is a 98 per cent Chinese society governed for the past 150 years by non-Chinese. It is primarily a Cantonese city, with significant communities of Hokkiens, Shanghainese, Chaozhou and Hakka people. Its culture and institutions reflect Cantonese and British models and great openness and sensitivity to international modernist influences. Hong Kong has made itself into the generally acknowledged capital of global Chinese (not just Cantonese) popular culture. (Taiwan's overseas cultural claims lie more in the area of higher Chinese culture.)

Macau is a Cantonese city-state, like Hong Kong now about to revert to Chinese rule. Unlike Hong Kong, its political influence has been Portuguese since the sixteenth century and its partly creolized society is shaped by Cantonese, Hong Kong and Portuguese influences.

Singapore is a city-state of dominantly but not exclusively Chinese population and the same kind of government. It is primarily Chinese because Chinese more than any other immigrants flocked to it when it was first established in 1819.[18] Its leaders must somehow cope politically and culturally with not only the significant presence of Malays and Indians but that of five varieties of Chinese. Mandarinization and the creation of a Singaporean identity that is heavily Chinese – but more than that – and focused on Singapore itself, is the method chosen. In passing, below the 'national' level we might mention another interesting example of early settlement and current majority status, that of Penang within Malaysia. Settled in the 1790s by Hokkiens it remains more Chinese than anything else, and mostly Hokkien Chinese.[19]

A second demographic type is the Large Minority Chinese society. Examples are Malaysia and Hawaii. In Malaysia, Chinese make up approximately one-third of the population – less than the Malays but considerably more than the Indians. Their presence in some specific places such as Malacca and Penang goes far back, but otherwise much of it dates from the nineteenth century. The history of Malaysia and the relative sizes of its three major ethnic groups have led to politics on a communal basis and Malaysia is the only place today where there are Chinese political parties as such. (In earlier decades of this century there were such parties representing the creolized Indonesian Chinese but these are now defunct.)[20] Hawaii occupies here a subnational status rather like that of Penang in Malaysia. Its Chinese population, almost all Cantonese from the one county of Zhongshan, does not constitute the majority in the Hawaii population but it is a significant minority though not functioning as such politically. The pattern of immigration and subsequent history of the Chinese of Hawaii is significantly different from that of other American Chinese.[21]

The third demographic type, Minority Chinese societies, is the most common variety – the one found everywhere else. These range from less than one per cent of the populations of the countries where they live to as much as five or ten per cent in some South-East Asian countries.

Although Chinese populations have historically been of importance in agricultural regions of many countries they have become increasingly urbanized since the beginning of this century. With a few exceptions, such as the Chinese in rural and small-town Malaysia, the trend has been towards concentrated residence in the capital cities or in secondary or regional urban complexes.

Migration destinations for Chinese, within China and overseas, have always tended to follow networks formed by kin and home-locality connections.[22] Accordingly, some countries of Chinese settlement may be dominated by a single ethno-linguistic subgroup of Chinese; or there may be two or more groups with one dominant or none dominant. Likewise, individual occupations often come to be monopolies or nearly so of given surname or home-locality mates. Which occupations are monopolized and by what group may vary from country to country and is not, in any case, ironclad or immutable.

Examples of subethnic concentration are: Houston, Texas, whose Chinese community is mostly made up of Taiwanese (Hokkien) engineers and other highly educated technical professionals, despite the fact that the American Chinese population remains predominantly Cantonese; Cantonese as the largest subethnic group in Vietnam, although historically the Chaozhou people controlled Saigon's rice market as they did those of Phnom Penh and Bangkok (where Chaozhou people dominate); and Manila's Chinese, dominantly Hokkiens, while many Cantonese there would rather live in Cantonese-dominant North America. The largest number of ethno-linguistic subgroups is found in Malaysia and Singapore, with no clear leader except on an occupational sector-by-sector basis.

The cultural adaptation of overseas Chinese has been shaped in part by cultural norms, models and experience brought from China. Most immigrants before 1949 were rural people in China, with some urban contact before migration and usually urban residence once established overseas. Since 1949, so much immigration has taken place after long residence in Hong Kong – or more recently also Taibei, Singapore or elsewhere – that rural–urban movement has been replaced by urban–urban transfer.

Chinese immigrants bring with them from China the generalized familial and interpersonal relationship norms associated with Confucianism. These include, in familial terms, well-defined roles for individual members, family solidarity and family discipline. More generally, these social norms stress the importance of personal networks and reciprocal assistance as the basis of action rather than more general and abstract principles or institutions. Education is prized for both its personally transforming qualities and its capacity for enhancing familial social status.

Besides these Confucian-derived norms there are values and practices derived from popular literature and from models of middle-class urban life as it developed in Late Imperial China. These include such practices as hard work and long-term family strategies for 'getting ahead'.[23] These models are most specifically derived from information about how Chinese migrants have adapted in urban situations within China and overseas.[24] Here the stock of information and models includes conceptions of how best to organize urban merchant activity, at both the level of the individual firm and that of the merchant community.[25]

Among small, entrepreneurial merchants several beliefs general to Chinese urban popular culture, and most salient in Fujian and Guangdong, are important. These focus upon values of loyalty and righteousness that should inform business and interpersonal dealings, but these values may also sanction the righteous use of violence against unethical authority figures and others. Overseas Chinese merchants' invocation of the hero-god, Guan Gong, has many meanings, most of them as yet unstudied.

Finally, there is the cluster of beliefs and models derived from impressions and memories of home-place. For many non-intellectual migrants, especially those not of middle or upper-class status, 'China' and 'Chinese culture' are most clearly associated with impressions of popular culture in home cities and towns, modified by more recent popular cultural outpourings from Hong Kong. Popular religious cults specific to Fujian and Guangdong have been taken everywhere overseas and are particularly important. But equally portable and important are Buddhist institutions and Chinese Christian churches – both Catholic and Protestant.

Another Fujian-Guangdong influence is the localized lineage. More than elsewhere in China, in these two provinces organizations made up of descendants of a common male ancestor – sometimes with memberships of several thousand persons – are of great importance. In these provinces they have historically been corporate bodies that hold land, operate schools, monopolize certain economic activities and provide for group defence.[26] Overseas, the knowledge of these organizations has two major effects. It encourages persons wherever they have settled to form common-surname (though not strictly kin-based) voluntary associations for mutual aid and other purposes. It also channels some remittance money to local projects in Fujian or Guangdong to benefit the overseas remitter's home lineage (but only after he has satisfied his remittance obligation to his immediate family).

Given this complex cultural baggage and their own ambitious, entrepreneurial, 'task-oriented' aggressiveness abroad, the overseas Chinese have usually encountered a mixed reception in their countries of settlement.[27] The Chinese have been thought to be important – perhaps essential – to the economic prosperity of the host country. But their cultural differences from local populations, their capacity for dominating or even monopolizing some economic sectors, and their seeming inclination merely to sojourn without commitment to the host country, have evoked fear, distrust and resentment.

These attitudes are common across both time and space. Their expression ranges from sixteenth century Manila to late twentieth century Vancouver.[28] Over a span of 400 years the Philippine local population has believed Chinese residents (or suspected sojourners) to be indispensable to the economy but culturally repugnant, economically overbearing and politically unreliable. Anxieties and resentments led to massacres of the Chinese in the seventeenth and eighteenth centuries, to expulsions and immigration restrictions and, more recently, to nationalist policies aimed specifically at the Chinese. For a period of 20 years up to 1975 Philippine Chinese were in the seemingly impossible position of being barred as non-citizens from several occupations while also barred by practice from becoming citizens. Yet they managed to adapt, evade, survive and even prosper.[29]

Across most of South-East Asia the Chinese have often been the target of restrictive or prohibitionist policies aimed at reducing their dominance of certain economic sectors and their ability to preserve their Chinese cultural heritage. Chinese schools, for instance, have often been closed down or tightly controlled. At the extreme, in Indonesia over the past quarter-century the open expression of Chinese culture, in any number of ways, has been illegal.[30]

Yet throughout most of South-East Asia intermarriage of Chinese with local populations has been substantial over the past century and even earlier. Indeed, in Thailand and the Philippines major portions of the political elite over many decades have themselves been the product of mixed Chinese–local intermarriage.[31]

By contrast, in North America intermarriage until quite recently has been fairly rare and dominant social groups have tended to keep the Chinese 'in their place'. That 'place', until recently, was a humble one. Unlike South-East Asia, where Chinese pioneered in many economic sectors and came to be pre-eminent in most, the early Chinese arrivals in North America found Europeans already in control of quite well developed economies and politically able to block Chinese opportunities by legislation or by discriminatory practice. The self-righteousness of European domination of North America is well illustrated in Ronald Takaki's recent *Strangers from a Different Shore*, where it is argued that in the United States (he could have included Canada) Americans of European derivation have always seen migrants from Europe as having a right to be in America but not so migrants from Asia, who were there only at the pleasure of Europeans and could be sent 'home' when no longer acceptable.[32] By law and practice both Americans and Canadians attempted to limit Chinese immigration and opportunity. Sometimes discrimination uniquely targeted the Chinese. In Canada only Chinese, of all immigrants, ever had to pay a $500 headtax to be admitted to the country (1903–23). Only Chinese, among all would-be immigrants, were ever totally excluded by law (1923–47).[33]

Altering white attitudes towards Chinese immigrants and their descendants has not been easy. In the past two decades American and Canadian immigration policies have become a mixture of family-reunion and business-professional programmes. Products of the former are often people of humble status and skills who fill jobs at the bottom of the economic hierarchy, unwanted by the majority of the population. Such immigrants may then try to work their way up, a practice accepted in North

American folklore and ideology. It is the other programme, attracting middle-class professionals and investors, that now causes problems for the white majority. The unaccustomed sight of Chinese coming in well up the socio-economic scale, driving expensive cars, buying large houses on arrival and otherwise acting as if they belong provides in Vancouver primal outcries of 'they're taking over!' – betraying anxiety that Europe's New World may now become anybody's New World.

Multiculturalism is official Canadian policy: it espouses integration, not an assimilating 'melting pot'. In the United States, too, multiculturalism has become the *de facto* policy though not officially stated. Yet both North American societies, whose self-image insists upon openness to immigrant opportunity, are in practice full of contradictions and ambivalences. Much of today's anxiety is about prosperous Chinese and their disproportionately large presence in North American universities. They are praised in the United States as a 'model minority', living the 'American Dream' of hard work, savings and upward mobility through effort and education; and are thus contrasted to other visible minorities who seem to lack these qualities and achievements. Yet there are fears that the Chinese, as usual, have 'overdone it', leaving too little, including continued dominance, for the dominant locals. They may indeed be 'taking over', by adapting so well to North American norms, and by spreading their culture to non-Chinese North Americans.[34]

Local attitudes and restrictions on Chinese economic opportunity have helped shape Chinese economic activities abroad; so have such influences as models of Chinese business organization and the existing structures and experiences of earlier Chinese settlers in the given locality. Commerce has usually been the preferred initial occupation. Going abroad to improve the family's economic position was seen as best served by getting into trade as soon as possible. Many Chinese who arrived as labourers could only go into commerce after saving enough to get free of their obligations. It might then be necessary to serve as a shop apprentice or frontier agent in a major Chinese merchant's distribution network before being able to realize the first step in the common overseas Chinese dream: owning one's business and being one's own boss. Where labourer's work was in a rural setting this process involved a rural-to-urban move – even when the 'urban' was merely a nearby town. In Peru, Cantonese plantation labourers left their plantations as soon as possible to open stores in towns.[35] In Hawaii the Zhongshan-county Cantonese who came to work on sugar plantations or other rural sites also gravitated to towns and to commerce.[36]

However, in many countries Chinese have been suburban market gardeners, feeding the nearby city, whether sixteenth-century Manila or twentieth-century Vancouver. The small-town or rural frontier Chinese operator of a general store is another common historical figure in the Philippines, the Canadian prairies or the Mississippi Delta. These were the lonely Chinese. They, and the Chinese laundry and restaurant operators in similar North American locations lacked the support of a Chinese community and became part of the larger society around them or retired eventually to more congenially Chinese surroundings. Not all life-time occupations have been commercial. Chinese artisans often dominated entire trades – commonly so in South-East Asia. In North America and Japan they found specific niches they could safely occupy, avoiding direct competition with more powerful local practitioners of the same art.

Merchants range from the small retail shopkeeper, through the distributor of imports and collector of exports, to the major wholesaler. Industrial development in Asia and relaxation of anti-Chinese discrimination in the West have led the Chinese into such fields as industrial production and management, and from that into service industries, banking and the media. Chinese-owned textile operations are now everywhere in Asia. In fast foods Manila's Jollibee chain, Chinese-owned, now competes successfully against McDonald's. Nearly all of Thailand's major banks are owned by people or groups of Chinese derivation. Most recently Philippine Chinese have occupied controlling positions in major English- (not Chinese) language newspapers in Manila. Important property developers in Asia, both within individual countries and internationally, are frequently of Chinese origin.

The general trends in business form and method among overseas Chinese have been from a Chinatown base outward to suburban bases; and from the intensive use of 'ethnic resources' to a mixture of Chinese and non-Chinese business creditors, suppliers, customers and other associates. The typical Chinese business at the start was either a partnership or a family enterprise. The increasing presence of women and children in twentieth-century overseas Chinese societies made possible substantial savings in operating such family businesses as grocery and convenience stores. It also raised problems about succession once the founder of the business died. Typically, family businesses have involved personal control by a dominant individual. With his disappearance and the growing size and complexity of a successful small firm, questions have been raised about how to make the transition to something more complex while retaining family control.

The opportunity structure for many overseas Chinese business people is shaped by a Chinatown business system. One gets a start there, and then either stays within it or, depending on the kind of business and existing opportunities, moves beyond it. How far and in what direction one can go will depend, besides resources, talent and fortune, on whether there are constraints on access to suitable education or training, free access to occupations and to citizenship and availability of credit. Chinatown credit is based heavily upon personal reputation or trust (*xinyong*) which, in turn, is based upon being known through business networks.[37]

Such networks and other kinds of social networking involve the organizations of overseas Chinese societies. The earliest of these in South-East Asia were either trade guilds or *huiguan* (common-home-locality associations), the two often being identical or overlapping in membership.[38] As in China, overseas migrants take home-locality as a prime basis for forming voluntary associations.[39] The scope of each association depends heavily upon the ethno-linguistic environment in the particular overseas locale. Where Chinese immigrants are from several provinces the divisions may be, at least initially, along provincial lines: a Fujianese association versus, say, a Cantonese one. A more common formulation, where there are immigrants from several parts of a given province, is to segment at the county level. At an extreme, in the Philippine case, the immigration from Fujian has been concentrated in so few counties as origin points that home-place associations in Manila are delineated at sub-county – even village – levels.[40]

A second kind of association is that based upon common surname, particularly important in North America but also found everywhere. Sometimes these will be more finely delineated – say, with two associations of a given surname, this having allegedly originated (in so far as its local bearers are concerned) in two different places in China. Or there may be surname alliance organizations in which the few bearers of a locally uncommon Chinese surname strengthen their capacity and influence by teaming up with members of a similarly deficient local surname group.

Local Chinese trade associations are somewhat different from traditional guilds. They are products of twentieth century China's development of Chambers of Commerce and their constituent interest bodies. In South-East Asia the trade association is important and the Chinese Chamber of Commerce which federates these associations and represents them also claims to represent all interests within local Chinese society. It thereby becomes the umbrella organization, the body that incorporates organized interest associations among the Chinese and speaks for them to the outside world while also trying to unify the Chinese population and control its affairs. South-East Asian Chinese Chambers of Commerce have extremely broad powers – economic, political and social – and they often control or supervise Chinese schools and Chinese social welfare work. In North

America trade associations and Chinese Chambers of Commerce have enjoyed indifferent success and the prime umbrella organization there has usually been the Chinese Benevolent Association, whose membership is made up of locality, surname and other voluntary associations.[41]

In the larger Chinese societies abroad there also exist a range of organizations that stress fraternity, politics and cultural preservation. These include the Hongmen bodies, sworn brotherhoods and certain kinds of musical associations. These organizations are concerned with the politics and culture (especially music and martial arts) of both China and Chinatown. They also honour Guan Gong and play mahjong. Some of the political parties take host country politics as their concern while others are oriented towards China's politics, and Chinatown's. The most conspicuous of the latter type is the Guomindang, founded early in this century in China and abroad and today the ruling party in Taiwan. Its influence varies from place to place, but it is most influential among either older and more conservative immigrants or the most recent arrivals from Taiwan. In some overseas locales, the Guomindang and the Hongmen (sometimes called the Chinese Freemasons) are major competitors in Chinatown politics.

It might seem that the existence of all these organizations implies considerable social fragmentation,[42] but this is not necessarily so, since membership is not exclusive and activists usually belong to several different kinds of organization. Among other things, these bodies provide several channels of social interaction, through which any individual member can communicate for reasons of business, kinship, home-locality or cultural objectives or Chinatown politics.

As business networks, membership (especially office-holding) in these organizations facilitates the flow of business information and the process of contacting business partners of all kinds. It also establishes reputation or trust *xinyong*, especially if one is an association officer and therefore to be taken seriously. For those Chinese businessmen who transcend the Chinatown business system this kind of reputation through 'ethnic resources' is of relatively little or no importance. Major property developers, for example, simply transact their business through banks and international corporate groups. But to the small entrepreneur these networks can be extremely valuable.

Association networks also extend overseas from a given Chinese society abroad, to other overseas Chinese societies, and to the home locality in China. The more 'globalized' Chinese migration and settlement have become, the larger in scale and more federative these associations have grown. Thus, there is now a world-wide Hakka association, world-wide associations of certain surname organizations and intra-Asian federations of certain county-based associations. For example, Jinjiang county in Fujian, the home county of most Philippine Chinese, is now part of an Asian federation of Jinjiang county associations, including ones in Manila, Singapore, Macau, Hong Kong and Taiwan.[43]

During the past two decades these associations, especially the home-locality ones, have been used to stimulate overseas Chinese investment in home-locality improvement and infrastructure. Although that has always been one of the functions of these associations, there seems now to be greater initiative coming from the home-locality itself.[44]

Such organizations, international and domestic to each migrant community, are not only for economic purposes. They were initially established to provide social support for new immigrants and in some parts of the world still do so. They also provide a social safety-net for immigrants of longer residence who find themselves short of money for trips to China, the costs of family illness or even their children's school fees. But in those Chinese migrant societies, where new immigration is slight, as in South-East Asia, these aspects of association work are now relatively less important and the business networking function relatively more so. It is probably the networking function of these 'traditional' associations that keeps them alive and proliferating in today's South-East Asia at a time when, according to predictions of social scientists and younger generation community activists, they should be disappearing.

Associations are also part of the politics of Chinatowns. They tend to become politically aligned, and leadership of umbrella organizations and the community as a whole is achieved by building up support through leadership in several key associations under the umbrella.[45] Traditional associations are also vehicles of cultural expression and retention. Their periodic rituals reinforce values and norms associated with kinship, respect for age and seniority, and attachment to home-place in China. Not surprisingly, these rituals and norms are of more interest to older than to younger, more acculturated Chinese. But the rituals bring together large parts of the Chinese society in some kind of 'community' form, even if only once or twice a year.

In addition to the traditional organizations, there have developed a number of new ones, especially during the past 50 years. These draw upon Western models and sometimes have Western affiliations; but their membership is Chinese and their activities and purposes reflect to some extent overseas Chinese interests. Christian church congregations are an obvious and very early example of organizations that draw on Western experience but are also expressions of several dimensions of modern Chinese culture. Alumni, or old-boy associations of overseas Chinese high-schools are another. Their members are usually business and professional people who raise funds for the school and engage in social activities in and out of Chinatown. Unlike the traditional associations, these bodies often are led by women and have women among their most active members. Other modern organizations for professionals and business people of modernist, cosmopolitan outlook may be associations devoted to the improvement of race relations or the promotion of Chinese culture among non-Chinese. (Examples include the Kaisa para sa Kaunlaran of Manila and the Chinese Cultural Centre of Vancouver.) Like the alumni associations, membership and leadership are both male and female. Finally, there are the Chinese Lions Clubs and other such fraternal-service organizations, which engage in community service and are vehicles for much business networking with both Chinese and non-Chinese.

Modern-style associations often have more contact and interaction with non-Chinese local populations than traditional Chinese organizations do. Some modern organizations exist primarily to serve residents of a given neighbourhood – Chinese and non-Chinese alike. Most kinds of both traditional and modern Chinese organizations however have gone in the other direction in recent years, forming international Chinese linkages which generate much exchange activity of a cultural, athletic and business nature with mainland China, Taiwan or Hong Kong, or with other overseas Chinese societies.

Since this exchange activity usually involves China or other Chinese societies the 'Chineseness' of a given recipient Chinese population may be reinforced thereby. A better reinforcer is the replenishing effect of new immigrants from China, Hong Kong or Taiwan. Overseas Chinese societies which are not replenished by new immigrants, and whose younger generation have ready access to most opportunities in the host country, are more likely to undergo an erosion of Chinese linguistic and cultural skills.

Canada is one example. In that country before World War II Chinese youth could expect few job opportunities outside Chinatown. These jobs required Chinese language and cultural skills, and Chinese schools teaching them were voluntary and supplementary to mainstream Canadian schools. Attendance at the 'after-school' Chinese school was due to parental pressure but students' absorption of their lessons also reflected their awareness of lost opportunity if they did not. As opportunities for Chinese youth opened up in post-war Canada, parents continued their pressure but young people began to think of these schools as irrelevant to their practical employment needs. Only later with the 'ethnic awareness' movement in North America during the 1960s and after did ethnic revival –

and growing trade with Asia – make Chinese language and cultural skills seem again to be 'practical', now both to Chinese and some non-Chinese.[46]

In the Philippines as elsewhere in South-East Asia the Chinese schools are comprehensive, full-day affairs, not 'after-school' supplements. By the 1970s young Chinese there had access to citizenship and thereby the full range of occupational opportunities. Since the schools teach both Chinese and other subjects student interest in the curriculum increasingly focused upon the latter to the detriment of learning in the former. Chinese parents in Manila, as elsewhere, were faced with a cruel educational choice: full economic opportunity was increasingly linked to university admissions and the standards for the latter escalated; there were not enough hours in the day to provide adequate instruction or study time for both Chinese and other subjects. Parents usually chose, reluctantly, to stress the non-Chinese subjects.

As more overseas Chinese students have entered universities everywhere and found greater acceptance in such environments, their social relations with non-Chinese have increased. Their lives after university have extended that multicultural environment as they have become professionals or business people in constant association with ethnically mixed colleagues, usually away from Chinatown.

These experiences since World War II have led to changing perspectives on 'Chineseness', seemingly different with each new generation. In a world of mobility, instant communication and multiculturalism, redefinitions go on at the level of the individual, the overseas Chinese 'community', the host nation and beyond. 'Chineseness' becomes not a matter of degree along a presumed acculturation scale with 'Chinese' at one end and totally assimilated 'Local' at the other. It becomes instead a question of kinds of Chineseness, generated at different times by changing blends of home and overseas influences. As increasing numbers of Chinese strike roots in the countries to which they have migrated, nationally-modified definitions become current and are popularly applied by governments that accept multiculturalism, both by overseas 'communities' and by ethnic Chinese individuals. The hyphenated Chinese appear: Chinese-Americans, Chinese-Filipinos and so on – expressions that imply integration, not assimilation and the maintenance of some forms of a Chinese culture alongside residential and political commitment to some non-Chinese national entity.

What does China think of all this? China participates in the neologizing. Since 1900 ethnic Chinese outside China have been called *huaqiao* ('overseas Chinese'), the clear implication being that they were sojourners and that China had some kind of claim upon them, emotional if no other. Since

World War II and especially in the post-1965 global diaspora and growing Chinese rootedness abroad, two new terms have appeared in China and overseas. The term *huayi* ('Chinese descendants') refers to the residue of earlier sojourner migration, usually to the generations that have emerged since World War II. The term *huaren* ('Chinese people') is neutral. It says nothing about generational status or cultural, residential or political commitments. It simply speaks of ethnic Chinese outside of China. All three terms are used in writings from mainland China, Taiwan and Hong Kong. They are also current among those overseas Chinese who still possess strong Chinese cultural skills, perspectives and orientations.[47]

CONTINUING RELATIONS WITH CHINA

In the late nineteenth century Chinese governments began to see the ethnic Chinese abroad as assets to China. There had long been an awareness of the local importance of their remittances to the economies of Fujian and Guangdong. To this was now added the belief that their earnings and skills were assets to China's defence and modernization. If Chinese abroad could be protected against the discrimination they so often faced by establishing Chinese embassies and consulates in the places of their residence they could not only continue their remittances but also be encouraged to invest in projects and causes beyond Fujian and Guangdong. By the 1890s China's politics also had overseas dimensions, as political competitors vied for overseas Chinese support. In the 1911 Revolution in China much of the financing and planning came from overseas Chinese, especially through the efforts of Sun Yat-sen.[48] Accordingly, Chinese abroad were given rights of representation in some Chinese government bodies. Shortly after 1900 the term *huaqiao* ('overseas Chinese'), with its connotations of temporary external residence and ultimate return to China, came into use.[49] The Citizenship Law of 1909 specified that any child of a Chinese father, whatever other nationality he might adopt, was a Chinese citizen. Chinese governments encouraged the founding of overseas Chinese schools and provided financial and curriculum aid, especially in the 1920s and 1930. When Sun Yat-sen's revived Guomindang came to power in China in the late 1920s it created a cabinet-level Overseas Chinese Affairs Commission and contacts between China and overseas Chinese became closer. Throughout the early twentieth century overseas Chinese, while maintaining their remittances to families and investments in local infrastructures in Fujian and Guangdong, supported national-level campaigns in China. The most notable of these were the resistance efforts against Japan in the early 1930s and especially during the Sino-Japanese War of 1937–45.[50] (Chinese in North America carried a double burden, supporting both China's resistance and the American and Canadian wars against Germany and Japan.)[51] Meanwhile, overseas Chinese were drawn into China's political struggles and civil warfare, partly by prompting from China and partly due to their own sojourning condition abroad.

The outcome of China's civil war in 1949 left two Chinas (mainland and Taiwan) vying for the support of the overseas Chinese. In mainland China special programmes and exceptional facilities were set up for returning overseas Chinese and their resident dependants. But these conflicted with revolutionary egalitarian policies and when the latter were in the ascendancy dependants and returnees found themselves in difficulty. In the late 1970s, as China opened up to more foreign involvement and began a rapid modernization drive, overseas Chinese were once again seen as an asset and new policies attempted to attract their investment and the transference of their skills. Taiwan lacked any overseas Chinese dependants but it had the resources to fund trips to Taiwan to encourage political support and investment and was better positioned than the mainland to aid overseas Chinese schools and other projects abroad.

Overseas Chinese societies were divided in their affiliations, these being influenced by the cold-war political affiliations of their countries of residence, the nature of Chinatown politics and many other considerations. In North America the Guomindang-related 'pro-Taiwan' factions dominated for 20 years following World War II. This was partly a development out of World War II,[52] but in the United States it also reflected the very close American cold-war ties with Taiwan. Only the rise of radicalism and 'ethnic awareness' in younger generations allowed the Chinatown left to surface and become politically competitive by the late 1960s. Here as elsewhere, Chinatown politics have ideological, symbolic and practical aspects, in which left and right members may gain business opportunities or Chinatown status by lining up with one side or another, even as both Chinas modified their claims and appeals.[53]

Meanwhile Hong Kong, neither asking nor offering any political affiliation or influence, was beginning to exercise cultural and economic influence as a kind of 'third China'. Before 1949 its life had depended upon an entrepot trading position in relation to China. Robbed of that by the cold-war boycott of mainland China, Hong Kong could survive only by developing in its own right as a small manufacturer and exporter, while retaining a certain access to the mainland. By the 1970s Hong Kong's economic success had made it not only a place to invest but a source of investment itself, the largest source, in fact, for post-1979 modernizing

China. It also exported talented emigrants around the world and a vibrant, unique form of modern Chinese popular culture that affected overseas Chinatowns everywhere.

That culture, along with the differing mainland and Taiwan versions of modern Chinese culture, have influenced the ongoing reassessments of 'Chineseness' abroad. These discussions are stimulated by the ever-growing number of Chinese abroad and their higher levels of educational and economic achievement. They are urged on by mainland China's modernization programme and its possible consequences; by anxieties about the future of Hong Kong after 1997; by the economic success of Taiwan and the other Newly Industrializing Economies, all bearing some influence of Chinese cultural norms;[54] by rapid changes in Chinese culture, in China and abroad; by the simultaneous existence of rooted and highly mobile overseas Chinese – the latter ready to move on to a third or fourth overseas residence, not necessarily back to China; by the complexity of contemporary overseas Chinese societies, caused by multiple migrations by poor and rich as well as by generational changes; and by the sense that in some way there is, or should be, some tie back to China. The discussion goes on in an atmosphere of diversity and fragmentation. Although the term 'community' is often used to refer to a given overseas Chinese local population, there is now less sense than before of a cultural consensus that might help sustain a community. Economic relations between overseas Chinese and China increase in importance. The home localities of overseas Chinese are now among the most prosperous and dynamic areas in China. Cultural exchanges go on constantly. But concern about trade with China is balanced by concern about the future of democracy and human rights in Hong Kong and mainland China. In countries where immigration continues newer immigrants, including some ethnic Chinese from Vietnam and Cambodia, form new Chinatowns, often suburban, as in Monterey Park, California, or the many Chinatowns of Metro Toronto. Global conferences are convened to discuss global overseas Chinese business and global Chinese culture.

In some discussions the term 'Greater China' is used. This seems to refer, most often, to the prospective combined economic power of mainland China, Taiwan, Hong Kong and Macau and perhaps the ethnic Chinese around the world. A second formulation, the work of intellectuals, is 'Cultural China', which includes ethnic Chinese wherever they are found. Like 'Greater China', 'Cultural China' may have economic and political dimensions and uses. The cultural discussion involves two groups: the large body of Chinese intellectuals now in diaspora and the growing well-educated middle-class business and professionals in overseas Chinese societies. Their version of 'Chineseness' stresses 'higher' culture, long associated with Confucianism. The intellectuals among them feel a special responsibility to redefine the Chinese heritage and culture in ways that will create a newer and better twenty-first-century China and to share that culture with non-Chinese for the benefit of all humanity.[55] Meanwhile the poor and small-business sector of the overseas populations also feels a pull and a responsibility towards China. But their version of Chinese culture is likely to be a popular one and their China is primarily the China of the home locality.

The history of overseas Chinese economic success and ambivalent local reaction to them continues. So does their own ambivalence about mobility and rootedness. The exact nature of a combined commitment to overseas residence and a sense of responsibility towards some kind of 'China' is not clear. Perhaps we can leave it at that by citing one version, a quotation in a recent publication of a Chinese home-locality organization in Hong Kong: 'Plant roots in Hong Kong; tie the heart to the home region.'

REFERENCES

1. *Huaqiao Jingji Nianjian* [Overseas Chinese Economy Yearbook], (Taibei: World Chinese Traders Convention General Liaison Office, 1989).
2. W.T. Rowe, *Hankow: Commerce and Society in a Chinese City, 1796–1889*, (Stanford: Stanford University Press, 1984), pp. 233ff.
3. R. Skeldon, *Population Mobility in Developing Countries: A Reinterpretation*, (London: Belhaven Press, 1990), pp. 125–6.
4. Wang Sing-wu, *The Organization of Chinese Emigration, 1848–1888* (San Francisco: Chinese Materials Center, 1978); Wang Gungwu, *China and the Chinese Overseas* (Singapore: Times Academic Press, 1991), pp. 3–8.
5. Chen Ta, *Emigrant Communities in South China* (New York: Institute of Pacific Relations, 1940).
6. Wang, *China and the Chinese Overseas*, pp. 4–6.
7. Cf. Wang, *China and the Chinese Overseas*, pp. 3–21.
8. E. Wickberg, 'The Chinese Mestizo in Philippine History', *Journal of Southeast Asian History*, 5 (1964) 62–100; G.W. Skinner, 'Change and Persistence in Chinese Culture Overseas: A Comparison of Thailand and Java', *Journal of the South Seas Society* 16 (1960) 86–100.
9. Yuen-Fong Woon, 'The Voluntary Sojourner among the Overseas Chinese: Myth or Reality', *Pacific Affairs* 56 no. 4 (Winter 1983–4), 673–90.
10. Wang, *China and the Chinese Overseas*, pp. 172ff; J.F. Warren, *Rickshaw Coolie: A People's History of Singapore (1880–1940)* (Singapore: Oxford University Press, 1986) p. 218; E. Wickberg et al. (ed.), *From China to Canada. A History of the Chinese Communities in Canada*, (Toronto: McClelland & Stewart, 1982) pp. 80–1, 94–5.
11. *Huaqiao Zhi–Zong Zhi* [General Treatise on Overseas Chinese] (Taibei: Huawai Qubanshe, 1956), p. 136.
12. J. Cushman and G. Wang (eds), *Changing Identities of the Southeast Asian Chinese Since World War II*, (Hong Kong: Hong Kong University Press,

1988); L.A.P. Gosling, 'Changing Chinese Identities in Southeast Asia: An Introductory Review', in L. Lim and L.A.P. Gosling (eds), *The Chinese in Southeast Asia* vol. II, (Singapore: Maruzen, 1983), pp. 1–14.
13. E. Wickberg, 'Notes on some Contemporary Social Organizations in Manila Chinese Society', in A.S.P. Baviera and T. Ang See (eds), *China Across the Seas: The Chinese as Filipinos*, (Quezon City: Philippine Association for Chinese Studies, 1992), pp. 49–50.
14. J.L. Watson, *Emigration and the Man Lineage*, (Berkeley: University of California Press, 1975).
15. Wickberg et al., *From China to Canada*, pp. 245–6.
16. M.H. Lai, *Cong Huaqiao dao Huaren* [From Chinese Sojourners to Ethnic Chinese] (Hong Kong: Joint Publishing, 1992), p. 526.
17. Wickberg et al., *From China to Canada*; Lai, *Cong Huaqiao dao Huaren*, pp. 417–532; Chen Hsiang-shui, *Chinatown No More: Taiwan Immigrants in Contemporary New York*, (Ithaca: Cornell University Press, 1992).
18. Wang, *China and the Chinese Overseas*, pp. 166ff.
19. Ibid., pp. 169ff.
20. L. Suryadinata, *Peranakan Chinese Politics in Java*, (Singapore: Singapore University Press, 1981).
21. C.E. Glick, *Sojourners and Settlers*, (Honolulu: University of Hawaii Press, 1980).
22. G.W. Skinner, 'Mobility Strategies in Late Imperial China: A Regional Systems Analysis', in C.A. Smith (ed.) *Regional Analysis*, vol. 1 (Economic Systems) (New York: Academic Press, 1976), pp. 327–64; F. Wakeman Jr. and Wen-hsin Yeh, 'Introduction', in Wakeman and Yeh (eds), *Shanghai Sojourners*, (Berkeley: University of California Center for Chinese Studies, Research Monograph no. 40, 1992), pp. 1–14.
23. D. Johnson, A. Nathan and E. Rawski (eds), *Popular Culture in Late Imperial China*, (Berkeley: University of California Press, 1985).
24. W.T. Rowe, *Hankow: Commerce and Society in a Chinese City, 1796–1889*, (Stanford: Stanford University Press, 1984) pp. 213–33; G.W. Skinner, 'Introduction: Urban Social Structure in Ch'ing China', in G.W. Skinner (ed.), *The City in Late Imperial China*, (Stanford: Stanford University Press, 1977), pp. 521ff; P. Golas, 'Early Ch'ing Guilds', in G.W. Skinner (ed.), *The City in Late Imperial China*, pp. 555ff.
25. Wang, *China and the Chinese Overseas*, pp.181–97, 281–4.
26. M. Freedman, *Chinese Lineage and Society: Fukien and Kwangtung*, (London: London School of Economics Monographs in Social Anthropology, no. 33, 1966).
27. Compare the reception given to overseas Indians: C. Clarke, C. Peach and S. Vertovec, 'Introduction to Part I: South Asians in Colonial and Post-colonial Contexts', in Clarke, Peach and Vertovec (eds), *South Asians Overseas: Migration and Ethnicity*, (Cambridge: Cambridge University Press, 1990), p. 37.
28. Wickberg, *The Chinese in Philippine Life, 1850–1898*, (New Haven: Yale University Press, 1965); Wickberg et al., *From China to Canada*.
29. Wickberg, 'Some Comparative Perspectives on Contemporary Chinese Ethnicity in the Philippines', *Asian Culture* [*Yazhou Wenhua*], 14 (April 1990), pp. 23–37.

30. L, Suryadinata, *Pribumi Indonesians, the Chinese Minority and China*, (Singapore: Heinemann Asia, 1978), pp. 153–64.
31. Skinner, 'Change and Persistence in Chinese Culture Overseas'; Wickberg, 'The Chinese Mestizo in Philippine History'.
32. R. Takaki, *Strangers from a Different Shore*, (Boston: Little, Brown, 1989).
33. Wickberg et al., *From China to Canada.*
34. Wickberg, 'Chinese Organizations and Ethnicity in Southeast Asia and North America since 1945: A Comparative Analysis', in Cushman and Wang (eds), *Changing Identities of the Southeast Asian Chinese since World War II*, p. 312.
35. B. Wong, 'The Chinese in Lima, Peru and New York City: A Comparative Study', *Comparative Studies in Society and History*, 20 (1978), pp. 351–2.
36. Glick, *Sojourners and Settlers.*
37. Lim and Gosling (eds), *The Chinese in Southeast Asia*, vol. I, pp. 1–130; D. Chen, 'Social and Economic Relations of Overseas Chinese Business', (PhD Dissertation, University of Oregon, 1976), University Microfilms 76–27, 633.
38. E. Armentrout Ma, 'Fellow-regional Associations in the Ch'ing Dynasty: Organizations in Flux for Mobile People. A Preliminary Survey', *Modern Asian Studies*, 18, no. 2 (1984), pp. 307–30.
39. G. Hamilton, 'Regional Associations and the Chinese City', *Comparative Studies in Society and History*, 21 (1979) pp. 338–53.
40. Chinben See [Shi Zhenmin], 'Feilubin Huaren Wenhua di Zhixu' [Persistence and Preservation of Chinese Culture in the Philippines], Zhongyang Yanjiu Yuan, *Minzuxue Yanjiu-sui Jikan* [Bulletin of the Institute of Ethnology, Academia Sinica, Taiwan], 42 (1976), pp. 119–206.
41. Wickberg, 'Chinese Organizations and Ethnicity in Southeast Asia and North America since 1945', pp. 304–7.
42. L. Crissman, 'The Segmentary Structure of Urban Overseas Chinese Communities', *Man*, 2, no. 2 (1967) pp. 185–204.
43. E. Sinn, 'Challenges and Responses: The Development of Hong Kong's Regional Associations, 1945–1990'. Unpublished paper presented at the 12th Conference, International Association Historians of Asia, 24–28 June 1991, p. 29.
44. Sinn, 'Challenges and Responses', pp. 24ff.
45. G.W. Skinner, *Leadership and Power in the Chinese Community of Thailand*, (Ithaca: Cornell University Press, 1958).
46. Wickberg, 'Chinese Organizations and Ethnicity in Southeast Asia and North America since 1945', p. 309.
47. Wang, *China and the Chinese Overseas*, pp. 3–40, 240–92.
48. Yen Ching Hwang, *The Overseas Chinese and the 1911 Revolution*, (Kuala Lumpur and London: Oxford University Press, 1976).
49. Wang, *China and the Chinese Overseas*, pp. 6–8, 240–6.
50. Ibid., pp. 198–200.
51. Wickberg et al., *From China to Canada*, ch. 14.
52. Ibid., pp. 227–8.
53. Wang, *China and the Chinese Overseas*, pp. 240–53.
54. Ibid., pp. 181ff, 222–35, 258–72.
55. The Living Tree' (A Special Issue on Chinese Culture) *Daedalus*, 120, no. 2 (Spring 1991).

2 Three Phases of South Asian Emigration

Ceri Peach

Although South Asia is the second most populous region of the world, its contribution to overseas migration is not great. Less than one per cent of its 1000 million people are found outside the immediate region. In 1987 estimates put the figure at 9.25 million (see Table 2.1) but this was before the 1991 Gulf War sent many fleeing back to the subcontinent. Overseas Chinese number three times as many at over 30 million, while Africans number 300 million overseas relative to 540 million in Africa itself, and there are 350 million overseas Europeans compared with 700 million in Europe.[1] South Asia has had long contacts with the outside world. Religious men and traders linked East Africa, China and Indonesia. However, the Indian subcontinent seems to have been a significant exporter of influence but not of population until the nineteenth century.

South Asian migration overseas has two main periods of direct outflow and a third period of secondary movement. The first period is that dominated by the Indenture movement. Indenture took place within the British imperial system but indenturing workers from British India was adopted in Dutch and French colonial territories as well. This period stretched from 1834 to 1920.

The second main phase is that of the post-war period of free market migration. This movement affected Britain in particular, but also the United States, Canada, some European countries and the Middle East.

The third movement overlaps the last and shares with it many of the destinations. This is the secondary migration, often forced, of the descendants of the first phase of migration. The most spectacular of these movements was the expulsion of Asians from East Africa, but there were also expulsions or repatriations from Burma and Sri Lanka and political flights of South Asians from Surinam and Fiji. East Indians (as South Asians are known in the Caribbean) were involved in the 'free market' movements from those areas as well.

THE FIRST PHASE: INDENTURE MIGRATION 1834–1920

Tinker[2] refers to the indenture system as 'a new system of slavery'. The timing of its inception, just after the 1833 emancipation of slaves in the

Table 2.1 Overseas South Asians by major area, by country and major region, late 1980s

UK	1 323 000	UAR	382 302
Netherlands	102 800	Oman	190 000
France	42 000	Kuwait	355 947
Germany	32 335	Yemen (PDR)	103 230
Spain	10 000	Iraq	85 000
Portugal	7 300	Saudi Arabia	79 987
Sweden	7 046	Qatar	51 500
Austria	3 131	Bahrein	48 050
Norway	2 900	Yemen	14 000
Switzerland	2 863	Lebanon	15 000
Denmark	2 552	Iran	6 300
Other	9 107	Jordan	4 506
		Other	319
Total Europe	1 545 034	Total M.E.	1 336 141
South Africa	839 000		
Mauritius	700 712	Trinidad	430 000
Kenya	70 000	Guyana	300 350
Reunion	65 000	Surinam	140 000
Tanzania	40 000	Jamaica	33 600
Libya	35 500	Guadeloupe	23 000
Malagasy	21 250	Martinique	10 000
Zambia	20 900	St Vincent	6 000
Mozambique	20 850	Grenada	4 000
Zimbabwe	16 000	St Lucia	3 840
Nigeria	14 000	Panama	2 100
Seychelles	5 200	Other	4 400
Malawi	5 000		
Liberia	3 066	Total Caribbean and	
Algeria	3 003	Latin America	957 290
Ethiopia	3 000		
Zaire	2 711	USA	500 000
Other	13 260	Canada	228 500
Total Africa	1 873 452	Total N.America	728 500
Malaysia	1 170 000	Fiji	839 340
Burma	330 000	Australia	99 200
Singapore	169 100	New Zealand	15 000
Bhutan	70 037	Other	569
Afghanistan	45 600		
Indonesia	30 000	Total Pacific	954 109
Hong Kong	20 180		
Philippines	12 100		
Thailand	6 802		
Brunei	5 500	GRAND TOTAL	9 262 282
Japan	2 685		
Other	752		
Total Asia	1 862 756		

Sources: A.H. Helweg, *Sikhs in England* (Delhi: OUP 2nd edn, 1986), pp. 214–20: C.G. Clarke, C. Peach and S. Vertovec (eds), *South Asians Overseas: Migration and Ethnicity* (Cambridge: CUP 1990), p. 2.

British Empire, is not accidental: indenture replaced direct slavery. In the West Indies, sugar plantations were still being established in Trinidad and in British Guiana in the 1830s. The Dutch, at a later date, were engaged in similar developments in their part of Guyana (Surinam) and the French in the islands of Martinique and Guadeloupe. Indenture was started in the French and Dutch colonies shortly after they abolished slavery. The French banned slavery in their colonies in 1848 and recruitment from their Indian colonies started in 1854 and from British India in 1861.[3] The Dutch banned slavery even later, 1863,[4] and started recruiting 'Hindustanis' in 1873.[5] The banning of slavery by the British government, coupled with the rejection by Afro-Caribbeans of plantation work, made the West Indian planters search for new labour sources. Similar searches took place in Natal, Mauritius, Fiji and elsewhere. Land-hungry India provided the source. Indentured migration must, therefore be seen as part of the imperialist and capitalist core-periphery relationship.

Table 2.2 outlines the volume of indentured labour received by the various overseas colonies of the British, French and Dutch; it also shows the 1980 estimated population of South Asian origin in these places. Mauritius received not only the largest Indian population (just under half a million between 1834 and 1912), but was probably the most 'Indianized' of the areas of settlement. They outnumber the Creoles of French-African descent by about two to one. In Guyana, where nearly a quarter of a million Indians settled between 1837 and 1917, they probably now constitute a majority of the population, although political manipulation has prevented them achieving political power except for a very brief period. A similar situation obtains in Fiji.

Not all of the migrant population during the indenture period went to indentured labour on the land. In East Africa and Malaya, they went as railway builders and as administrators. In South Africa, they worked in the coal mines as well as on sugar and cotton plantations. Nor was indenture the only scheme of recruitment. Under the *kangani* and *maistri* systems, an Indian returning home was commissioned as a middleman to recruit labourers from his home district.[6] The *kangani* system of labour recruitment was particularly important and replaced indenture in Malaya after that system was abolished there in 1910.[7] There was also considerable movement of free migrants, traders, doctors and lawyers ('passenger Indians') to areas where South Asian communities were settled. Seamen had been important migrants since the days of the East India Company and Lascar sailors had formed the nucleus of Bengali settlement in the East End of London. Finally, one should not overlook the importance of student movement and the settlement of scientists, doctors and academics in Britain.

Table 2.2 Indentured Indian immigration by colony in the nineteenth and early twentieth centuries, and by Indian population by country, 1980

Colony/Country	*Period*	*Indian immigrants*	*Indian population (estimate) 1980*
Mauritius	1834–1912	453 063	623 000
British Guiana (Guyana)	1838–1917	238 909	424 400
Natal (South Africa)	1860–1911	152 184	750 000
Trinidad	1845–1917	143 939	421 000
Reunion	1829–1924	118 000	125 000
Fiji	1879–1916	60 969	300 700
Guadeloupe	1854–1885	42 326	23 165
East Africa (Kenya/Uganda)	1895–1901	39 771	79 000/430
Jamaica	1854–1885	36 420	50 300
Surinam (Dutch Guiana)	1873–1916	34 000	124 900
Martinique	1854–1889	25 509	16 450
Seychelles	1899–1916	6 319	n/a
St Lucia	1858–1895	4 350	3 700
Grenada	1856–1885	320	3 900
St Vincent	1861–1880	2 472	5 000

Source: Clarke, Peach and Vertovec (eds), *South Asians Overseas*, p. 9.

The Nature of Indenture

Indenture generally required the emigrant to apprentice himself or herself to a master for five years, although a range from three to ten years is found. There were often additional conditions. In the case of Trinidad, the worker was indentured for five years but was obliged to work for a further five before he/she was eligible for a free return passage.[8] At the end of this period he could either retire, passage paid, to his homeland or else take a piece of land to cultivate as his own in the overseas territory. These contractual promises were not always fulfilled.[9] The Protectors, who were appointed by the Indian government, seem to have been fairly powerless. For whatever reasons, the return flow of formerly indentured workers was generally small. For example, of the 143 939 Indians who migrated to Trinidad under indenture, only 33 294 eventually returned to India.[10] Clarke reports that not an inconsiderable number of repatriated indentured migrants signed on for further periods in Trinidad.[11] However, in Martinique and Guadeloupe, just under a third of the immigrants seem to have been repatriated[12] and a similar rate is true of Surinam.

The agreements between governments which controlled these contracts required women to be included in the indentured group in the ratio of 4:10.[13] Women also seem to have had a shorter period of indenture – three rather than five years.[14] The communities which resulted from indenture thus tended initially to have a skewed sex-ratio. 'Passenger Indians', on the other hand, tended to have a much more balanced structure so that in South Africa, for example, although passenger Indians constituted only 10 per cent of the migrants, their descendants now form 30 per cent of the current South Asian population in that country.[15] The evidence for this is that indentured Indians were largely Tamil or Telugu-speaking Hindus from the south of the country, while passenger migrants were largely Gujarati, Urdu or Hindi-speaking Muslims from the north[16] so that the differences between the proportions of the groups at the beginning of immigration and the present time indicates the different rates of natural growth.

Impact of Indenture on Receiving Societies

The number of migrants to the indenture territories differed quite considerably, as did the percentage of the population that they constituted in those territories. In Mauritius, Fiji and Guyana they came to constitute a majority, but only in Mauritius has that demographic fact been turned into democratic political control. In Fiji, the constitution created by the British on independence, guaranteed native Fijian control. Coups in 1988[17] and subsequently have kept the South Asians out of power. In Guyana, after a brief period of power under Cheddi Jagan in the late 1950s and early 1960s, British and American machinations against his Marxist-inspired government ousted the Indian leadership and encouraged politics to split along ethnic rather than class lines.[18]

In Guyana and in Trinidad, the East Indians were, in terms of esteem, at the bottom of the structure. Slavery had stigmatized work on the land. East Indians were rural land-workers. In Sri Lanka, their position was even poorer. Whereas in Guyana and Trinidad, East Indians had citizenship and had achieved landownership and some degree of upward mobility through education, in Sri Lanka they were stateless estate labourers.[19]

In African countries where they had settled, South Asians, on the other hand, formed a middle-class elite between the indigenous African population and the white rulers. In South Africa, whites tried at first to disenfranchise them and later (in the 1980s) to co-opt them to bolster the white position against the African population.[20] In East Africa, the removal of white colonial rule left them in an exposed position. Non-citizens were first forced out of civil service occupations in Kenya and Uganda. Later, in

Uganda they were forced out of the country itself. South Asians also formed a middle class in Burma during British rule, but independence brought ethnically exclusionary policies and forced many of them (the skilled service occupation sector in particular) out of the country. Only the poor South Asians continued to live there, keeping a low profile, adopting Burmese dress and language in public.[21] Even here, however, Muslims along the border with India and Bangladesh were being forced out of the country in 1992. In the remainder of South-East Asia, in Malaysia and Singapore, South Asians also form a low profile community: in Malaysia they are tolerated because they are perceived as less of a threat to the Malays than the Chinese; in Singapore, they help to boost the multicultural image of what is, in effect, an ethnic Chinese state.

Regional Origins

It is not clear to what extent the origins of the overseas South Asians reflect the balance of the population of the subcontinent. In the subcontinent, about three-quarters come from the Aryan north and about a quarter from the Dravidian south. Northerners (Punjabis and Gujaratis) are dominant in Britain, Canada and the British Caribbean. Tamils are prominent in the settlement of Malaysia, Singapore,[22] and also in the French territories (because of the Pondicherri connection). Gujaratis and Punjabis were prominent in the East and South African settlements. Northerners, from Uttar Pradesh in particular, predominate among the Trinidad East Indians.[23] However, Bengalis seem curiously under-represented in the overseas movements. Given that Calcutta was the main port of India for a long time, that Bengal was desperately overpopulated, and that there was a tradition of sailing on British merchant ships, surprisingly few Bengalis migrated.

Religious Composition

In religious terms, the overseas population is possibly less representative. Muslims and Sikhs are probably over-represented relative to Hindus, but their proportions differ in different countries. For example, Sikhs form about 2 per cent of the Indian population but probably about 11 per cent of the East African Asian population (see Table 2.3) and a quarter of the Indian ethnic population of Britain. Muslims form about a third of the population of the subcontinent, but just under half of the South Asian ethnic population in Britain. In Trinidad, Muslims seem to have formed less than 15 per cent of the indentured population.[24] It is difficult to know the position with any certainty for small groups such as Jains and Parsis.

Likewise, the position of Christians is not clear although they seem over-represented among the migrants to Britain (Table 2.3). In addition, significant conversions have taken place among the East Indian communities in the Caribbean. Distance from the subcontinent of the areas of settlement had quite profound effects. In many of the East African countries and Burma, proximity was sufficiently close for some migrants to leave their families in the subcontinent and not have family settlement. Even if families were settled, distances were not so great as to prevent return to the homeland in the search for brides or grooms for their children. Under these circumstances, the maintenance of traditional values and practices was relatively easy, particularly for the richer classes.

In the Caribbean, on the other hand, social renewal through arranged marriages with the homeland was not a practical proposition. This led to greater reduction in traditional caste divisions, if marriage was to be kept within the South Asian population. Distance was not the only factor in this consideration. The size of the community was important also. With relatively small communities at considerable distance from the subcontinent, adoption of English as the lingua franca (or French in the French territories) became more likely. In effect, there was a tendency for small isolated communities to become more homogeneous. Clarke[25] has shown that marrying out of religion was reasonably common for East Indians in San Fernando, Trinidad, but that marrying across the racial divide was much less common.

Caste

Caste is thought by some to have become profoundly changed or to have disintegrated in almost all of these proletarian overseas communities, while its influence has remained much stronger in urbanized and middle-class societies. The absence of precise occupational specialization, precise ritual roles, and inter-caste dependence in the new setting, the confusion of purity and pollution notions due to caste co-residence and commensality on estates[26] must have contributed to the reformulation of subcastes into larger groupings. Sceptics have pointed out that more Brahmins seem to have disembarked in Port of Spain than embarked in Calcutta. High caste status certainly survived in Trinidad as is evidenced by Naipaul in *The Middle Passage*. However, among the urban-based merchants and civil servants in East and South Africa, caste consciousness remained high even though the caste-based interaction and exchange ceased to function. Where direct migrants from the subcontinent meet migrants from secondary dispersal from indenture societies, the differences in caste practice become apparent

Table 2.3 South Asians in Great Britain: Religions by country of origin, 1982

Religion	*All Asian*	*Indian*	*Pakistani*	*Bangladeshi*	*African Asian*
			per cent		
Muslim	46	16	96	96	24
Sikh	20	43	–	–	11
Hindu	27	31	1	2	60
C of E	2	3	1	1	–
R C	3	4	1	–	3
Other	1	1	–	–	–
None	1	1	1	1	–
No answer/ Don't know	1	1	1	1	2

Source: Brown, *Black and White Britain*, p. 24.

and secondary migrant communities have difficulty in gaining acceptance from ritually 'purer' direct movers from the subcontinent.[27]

SECONDARY DISPERSAL

If colonialism brought into existence the indenture colonies of South Asians overseas, decolonization contributed substantially to their secondary dispersal. Decolonization unleashed the xenophobic forces of the indigenous populations into which the immigrant indentured workers and passenger migrants had been introduced.

Sri Lanka and Burma

In some cases, Indians were unable to take up citizenship of the newly independent countries; in other cases, they were unwilling to do so. In Sri Lanka, for example, most of the Indian Tamils resident in the country were made non-citizens by the citizenship laws introduced shortly after Independence. In 1964 about 134 000 Indian Tamils, less than 10 per cent of the total, had gained Sri Lankan citizenship. In 1964, by the terms of an agreement reached between Sri Lanka and India, it was decided that 525 000 'stateless' persons of Indian origin in Sri Lanka would be repatriated to India and 300 000 persons would be given citizenship in Sri Lanka, over a period of 15 years. The position of the remaining 150 000 was left undecided at the time. In 1974 another agreement was signed between the two countries, according to which, half the remaining number would be given Indian citizenship and the other Sri Lanka citizenship.[28]

The extent to which these agreements have been carried out is unclear, but between 1964 and 1969 only 25 000 Indians left Sri Lanka for India.[29] Tinker, writing in 1977,[30] indicated that Indian Tamils still constituted 10 per cent of the Sri Lankan population, no different from the proportion at Independence. This would suggest that, with continued population growth, Indian Tamils may now number 1.5 million. In fact, with the civil war in Sri Lanka during the 1980s, there were probably more Ceylon Tamils who fled to India than there were Indian Tamils who were repatriated. At the beginning of 1992, there were estimated to be about 200 000 refugee Ceylon Tamils in India.[31]

In Burma, resentment against the Indian migrants and settlers seems to have been of long standing. Although Indians formed only 8 per cent of the total work force in Burma in the 1930s, they were concentrated in key positions which made them appear conspicuous.[32] When the Japanese invaded in 1941, about half a million Indians fled the country. Both the British and the Burmese resisted Indian attempts to return to Burma after the war,[33] and with Burmese Independence, there began the familiar process of restricting Indian access to citizenship, reduction of their economic livelihoods and finally expulsion after 1962 when General Ne Win seized power. After Independence in 1948, there were about 800 000 Indian and 100 000 Pakistani resident aliens in Burma. By 1964, 300 000 Indians had left for India as refugees.[34] Those who fled were the elite. About 250 000 Indians were left in Burma, often adopting Burmese dress and names, living on the fringes of Burmese society and often stateless.[35]

East Africa

The two most dramatic examples of flight are rather contrasting: they are the East African Asians and the Surinamese. With Independence in Kenya and Uganda came a process of Africanization. Indians formed the middle-class entrepreneurial elite in these countries. On Independence, they were able to acquire citizenship, but there was a considerable reluctance to do so, except among the Ismaili minority. This 15 000 strong minority in Uganda did so *en masse*.[36] At Independence in 1962 there were 77 400 Asians in Uganda. Between 1969 and 1971 their numbers fell to 50 000. After Idi Amin staged a coup and came to power in 1971, he accused the Asians of being socially and economically exclusive (failing to intermarry and controlling the economy). He informed the governments of Britain, India, Pakistan and Bangladesh that they would have to take back all of their nationals, the first plane-load of expellees arriving in Britain in 1972. As Amin's ruthlessness grew, even the Ismailis who held Ugandan citizenship found that their claims were not recognized and fled the country.[37]

The Kenyan and Tanzanian exodus shared many of the same elements as those from Uganda, but was less extreme. British fears of an East African Asian flight precipitated a self-fulfilling prophesy. Asians hurried in 1968 as West Indians had hurried in 1962, to 'beat the ban' of the impending Commonwealth Immigrants Act.[38] By 1990, South Asian communities in Kenya and Tanzania were about a quarter of the size that they were in the early 1960s.[39] Ghai argues that in a race to keep up with the pace-making Ismaili, who pioneered communal facilities such as schools, burial places and recreational facilities, other ethnic, religious or caste groups did the same and, in the end, succeeded in fragmenting the South Asian population in East Africa.[40] British policies of preventing Asian land purchase and keeping them from rural trade made them even more obviously an alien urban group,[41] and stoked up the fires of African nationalist resentment which scorched them after independence.

Bhachu [42] has shown how the twice migrant Sikhs, forced out of East Africa, differed significantly from direct migrants from the subcontinent in their orientation in Britain. Critically, India was not 'home' for them; they lacked the myth of return.[43] A fascinating literature is developing which shows the somewhat tense interaction of twice migrants with single migrants in a third country. Bhachu shows it for East African Sikhs in Britain; Vertovec shows it for Trinidadian Hindus in London.[44]

Surinam

Between 1873 and 1917, 34 024 Indian immigrants were conveyed to Surinam.[45] They were recruited mainly in UP and Bihar[46] and shipped via Calcutta. About a third seem to have returned to India at the end of their indenture. The same skewed sex-ratio that characterized Indentured migration elsewhere was found in Surinam. Although most of the returnees were male, even in 1939 there were only 827 women to every 1000 men. The Dutch also recruited from their own colonies in Indonesia. Between 1891 and 1938, 32 620 Javanese came in of whom just over 20 per cent returned.[47]

The Surinamese exodus differs substantially from that of East Africa. The Surinam ethnic position was much closer to that of Guyana or Trinidad than to East Africa. Indians and Javanese were at the bottom of Surinamese society and although they held political power in the early 1970s they were ousted by the Creoles. Fear of Creole domination and the loss of Dutch citizenship after Independence in 1975, precipitated a huge flight to the Netherlands before the effective date.

By 1964 the 'Hindustanis' (as South Asians were known) were just slightly less numerous than the Creoles: 38.5 per cent as against 39.3 per cent; the Indonesians numbered a further 16.6 per cent.[48] However, political power lay with the more urbanized Creole population. The prospect of Creole domination precipitated a panic flight which affected the Hindustani and Indonesian population disproportionately. Between 1970 and 1975, 80 583 Surinamese migrated to the Netherlands.[49] This is the equivalent of about a quarter of its 1980 population. Indeed, the United Nations Centre for Human Settlements[50] indicates a decrease in Surinam's population between 1970 and 1980. Cross and Entzinger estimated that about 40 per cent of the 200 000 Surinamese present in the Netherlands in the mid 1980s were Hindustani,[51] about 80 000. Subsequent coups have precipitated further flights of population. Between 1970 and 1980 the Hindu population of Surinam is thought to have decreased from 60 000 to 10 000.[52]

ECONOMIC MIGRANTS IN THE PERIOD SINCE 1945

Beginning just before the secondary movement, but overlapping it, is the second major movement from the subcontinent. The migration has taken place largely since 1947. These movements have been concentrated in areas strongly affected by the upheavals of the 1947 Partition, but the causes of movement are not directly related to that event. It is mainly a movement to the industrialized world and particularly to the UK, USA,[53] Canada,[54] and Australia.[55] At the beginning of the 1990s, there were about 1.5 million South Asians living in Europe, about 0.75 million living in the USA and Canada and over 100 000 living in Australia and New Zealand. However, since 1973, there has also been a major movement to the oil-rich countries of the Gulf. At the beginning of 1990, this population numbered about 1.4 million (about the same size as that living in the much larger population of Europe) but the 1991 Gulf War produced a major upheaval. The effects of flight and reflux are not yet clear but some indication of their scale is given in a later section of this chapter.

By far the most important single destination for South Asians living in the developed world is the United Kingdom. Of the 1.5 million living in Europe, 1.2 million live in the UK. The USA with five times the population of the UK has only about one-third of the number of South Asians. The reasons for this concentration are bound up with the post-war labour shortage in Britain and the Commonwealth membership of the successor states of British India. This made the citizens of these countries simultaneously citizens of the United Kingdom. Up to the enactment of the British 1962 Commonwealth Immigrants Act, this gave them the right of entry and settlement in the UK. The 1962 Act controlled worker immigration and successive legislation further restricted the access of dependants.[56]

However, the movement to Britain should not be seen as a fortuitous outcome of a historical colonial tie. The British demand for labour was a particular example of a general Western European phenomenon. Germany, France and Britain (as well as the Netherlands, Belgium, Luxembourg, and Switzerland) sucked in labour and dependants. Even the Socialist countries of Eastern Europe had miniature *Gastarbeiter* labour movements from Vietnam, Angola and Mozambique. The French and Germans divided the Mediterranean fringe between them. France drew a foreign population of four million from Italy, Spain and Portugal and its former North African territories of Algeria, Morocco and Tunisia, and also drew population from sub-Saharan Africa. Germany drew its four million *Gastarbeiter* and dependent population from the southern and eastern quadrant, from Italy, Yugoslavia and Greece, but above all from Turkey. Austria and Switzerland showed a rather 'German' pattern in their foreign workers while Belgium showed more of a 'French' pattern. The Dutch and the British, however, showed a far more colonial dependency pattern of source areas, the British movements being dominated by the Irish, the Caribbeans and the South Asians.

West Indian movement to Britain was marked by a very close adjustment between British economic cycles and migration.[57] The movement began effectively in 1948, peaked in 1961 and was over by 1973. The correlation coefficient for net West Indian immigration against British unemployment was $r = -0.65$ for the period 1955–74.[58]

Movement from South Asia was tied less closely to British economic conditions. The movement started about eight years later than that from the West Indies and was more disrupted by British political legislation. The correlation coefficient for the relationship between British economic indicators and Indian and Pakistani immigration is neither as large nor as significant as that for the West Indian movement[59] but does, nevertheless, show the importance of economic motivation. The movement from India slightly preceded that from Pakistan, while that from Bangladesh was even further behind. The peak period of arrival for West Indian-born persons living in Britain in 1987 was 1955–64,[60] for the Indian- and Pakistani-born it was 1965–74 and for the Bangladeshi-born 1980–84. The gross movement from Bangladesh, perversely, shows a direct rather than a negative correlation with unemployment in Britain.[61]

It has been said that West Indians in Britain face an 'Irish' future while Asians face a 'Jewish' future. By this it is implied that West Indians will remain concentrated in blue-collar occupations, council housing and with low educational qualifications, while Asians will move more to entrepreneurial activities and self-employment in white-collar jobs with owner occupation and higher educational certification. There is some justification for this generalization. The Indian male occupational structure in Great Britain in 1987/89 was very close to the white population and above it at the top end. For example, 11 per cent of Indian men were in professional occupations compared with 7 per cent for the white population; 19 per cent were employers or managers compared with 20 per cent for the white population and 16 per cent were in other non-manual occupations compared with 18 per cent for the white population.[62] The comparable figures for the (regrettably) combined Pakistani/Bangladeshi population were 5 per cent, 16 per cent and 11 per cent respectively.

However, as with most generalizations, this hides many contrary examples. Perhaps, as far as the Asian population is concerned, the most serious qualification is that it hides major differences in housing tenure characteristics between the national origin groups. The Indians and Pakistanis have the highest housing tenure profile while the Bangladeshis are, in some ways, closer to the Caribbean model. Indian households were 77 per cent owner occupiers in Great Britain in 1983–5;[63] Pakistani households were 74 per cent owner occupiers while the proportion for the white population was 59 per cent and that for Bangladeshi households, 35 per cent.

The most recent (1987–9) estimates of the South Asian ethnic population of Britain is that Indians number 779 000, Pakistanis 433 000 and Bangladeshis 111 000.[64] These figures include direct migrants from the subcontinent, secondary migrants from third countries, mainly East African, together with children born in the UK (538 000 out of the total ethnic population of 1 323 000 from the three countries had been born in the UK).[65]

The most recent available figures which disaggregate the proportion of children born in the UK to individual national origin groups, refer to 1985. At that time, 241 000 out of the then total Indian ethnic population of 689 000 (35 per cent) had been born in the UK. For the Pakistani population 177 000 out of 406 000 (44 per cent) were UK born, while for the Bangladeshi population, 28 000 out of 99 000 (28 per cent) were UK born. Of the Indian ethnic population 39 per cent had been born in India and 20 per cent in East Africa. About one per cent of the Pakistani ethnic population was East African born. Most East African Asians are classified as 'Indian' and roughly 20 per cent of the Indian ethnic population of the UK is East African in origin.

Our best estimates of the religious origins of the South Asian population living in Britain come from C. Brown's work.[66] These show that just under half (46 per cent) of the South Asian population living in Great Britain in 1982 was Muslim, 20 per cent was Sikh and 27 per cent was Hindu (see Table 2.3).

Northern languages, Punjabi, Urdu and Gujarati, also predominated with Bengali growing in importance as Bangladeshi migration expanded. Using figures provided by K. Knott[67] it seems that, very approximately, about 54 per cent of the South Asian population in Britain is Punjabi-speaking, about 27 per cent Gujarati-speaking, about 8 per cent Bengali and about 4 per cent Hindi. (The Gujarati and Hindi proportions are the least reliable of these estimates).

The Gulf

One of the effects of the massive redistribution of world wealth in the aftermath of the 1973 petroleum price rise, was the huge increase in demand for labour in the Gulf.[68] The South Asian population in the Middle Eastern oil-rich countries grew from about half a million in 1975 to nearly two and a half million in 1985.[69] The Gulf crisis from August 1990 to early 1991 produced a major upheaval for contract workers from India, Pakistan and Bangladesh in Iraq and Kuwait. At the beginning of 1990, there were about 366 000 such workers in the two countries. Between August and December 1990, however, the number plummeted as 322 000 returned [70] (Table 2.4).

The impact on remittances from Iraq and Kuwait can be judged from the fact that in 1987, Bangladeshis remitted US$138 million from these two countries, Pakistanis remitted US$307 million and in 1985 Indians US$203 million.[71]

CONCLUSION

South Asians seem oppressed wherever they have settled overseas. In Burma, they were criticized because they married Burmese women; in Uganda and Tanzania they were criticized because they did not marry African men.[72] In South Africa they were ground between the lower millstone of African economic envy and the upper millstone of white racial discrimination. In Fiji, Guyana and Surinam their demographic size was negated by political rape. Where South Asian societies have become large or wealthy, fissiparous tendencies have developed. There

Table 2.4 Returnees from Kuwait and Iraq to South Asian countries from 2 August to early December 1990

Country of origin	*Contract workers*	*Returnees*
Bangladesh	90 000	62 000
India	181 000	170 000
Pakistan	95 000	90 000
Sri Lanka	101 000	73 000

Source: G. Battistella, 'Gulf Returnees: Profile of a Crisis', *Asian Migrant*, 4, 1 (1991), p. 25.

seems to be a critical threshold above which one moves from a South Asian solidarity into a series of ethnic particularities. Many of the rivalries from the subcontinent have been transplanted into the societies and the political issues which excite passion are often those of the homeland rather than those of the new home. The issue of Khalistan looms large for the overseas Sikhs, as does Indian control of Kashmir for many Kashmiris and Mirpuris.

However, one could also distinguish between optimistic and pessimistic patterns of settlement. The contrast between the 'Irish' and 'Jewish' futures referred to in relation to Caribbean and South Asian settlement in Britain could be extended to contrast the position of some areas of overseas settlement. For example, the indenture settlements of Trinidad, Guyana and Surinam as well as the Gulf could be classified as 'Irish'. They are dominated by manual labour and often primary activities. The pattern of East Africa, the USA, Canada and Australia, on the other hand, is more 'Jewish'. It is the settlement of skilled technical workers: white-collar professionals. The British pattern is ambivalent. The Indians and Pakistanis show a 'Jewish' pattern while the Bangladeshis seem currently to be moving in an 'Irish' trajectory.

REFERENCES

1. C.G. Clarke, C. Peach and S. Vertovec (eds), *South Asians Overseas: Migration and Ethnicity* (Cambridge: CUP, 1990), p. 1.
2. H. Tinker, *A New System of Slavery: The Export of Indian Labour Overseas 1830–1920* (London: OUP, 1974).
3. Singaravélou, 'Indians in French Overseas Departments: Guadeloupe and Martinique' in Clarke, Peach and Vertovec (eds), *South Asians Overseas*, pp. 75–7.

4. R. Cohen, *The New Helots: Migrants in the International Division of Labour* (Aldershot: Gower 1988), p. 7.
5. R.A.J. van Lier, *Frontier Society: A Social Analysis of the History of Surinam* (The Hague: Martinus Nijhoff, 1971), p. 217.
6. H. Tinker, 'Indians in Southeast Asia: Imperial Auxiliaries' in Clarke, Peach and Vertovec (eds), *South Asians Overseas*, p. 48; S. Vertovec, 'Oil Loan and recession in Trinidad Indian villages', in ibid., p. 108.
7. Tinker, 'Indians in Southeast Asia: Imperial Auxiliaries', ibid., p. 48.
8. C.G. Clarke, *East Indians in a West Indian Town: San Fernando, Trinidad, 1930–70* (London: Allen & Unwin, 1986), p. 10.
9. A. Lemon, 'The political position of Indians in South Africa', in Clarke, Peach and Vertovec (eds), *South Asians Overseas*, p. 131.
10. Clarke, *East Indians in a West Indian Town*, p. 9.
11. Ibid., p. 10.
12. Singaravélou 'Indians in French Overseas Departments: Guadeloupe and Martinique' in Clarke, Peach and Vertovec (eds), *South Asians Overseas*, p. 76.
13. H. Tinker, *The Banyan Tree: Overseas Emigrants from India, Pakistan and Bangladesh* (Oxford: OUP, 1977), p. 4.
14. Clarke, *East Indians in a West Indian Town*, p. 12.
15. Lemon, 'The political position of Indians', pp. 131–2.
16. *Official Yearbook of the Republic of South Africa* (Johannesburg, 1982), p. 79.
17. V. Lal, 'The Fiji Indians: Marooned at home', in Clarke, Peach and Vertovec (eds), *South Asians Overseas*, pp. 113–15.
18. Tinker, *The Banyan Tree*, pp. 66–73.
19. H. Balakrishnan and H.M. Gunasekera 'A review of demographic trends', in K.M. de Silva (ed.), *Sri Lanka: A Survey* (London, 1977), p. 126, fn. 12.
20. Lemon, 'The political position of Indians', pp. 131–48.
21. Tinker, 'Indians in Southeast Asia', pp. 45–6.
22. Tinker, *The Banyan Tree* pp. 97–8; Tinker, 'Indians in Southeast Asia', pp. 46–50.
23. Clarke, *East Indians in a West Indian Town*, p. 9.
24. Clarke, ibid., p. 10.
25. Clarke, 'Residential segregation and intermarriage in San Fernando, Trinidad', *The Geographical Review* 61, 2 (1971), 198–218.
26. Clarke, Peach and Vertovec (eds), *South Asians Overseas*, p. 13.
27. S. Vertovec, 'Community and congregation in London Hindu temples: divergent trends', *New Community* 18, 2 (1992), pp. 251–64; S. Vertovec, 'Caught in an ethnic quandary: Indo-Caribbean Hindus in London' in R. Ballard (ed.), *Desh Pardesh: The South Asian Presence in Britain* (forthcoming).
28. Balakrishnan and Gunasekera, 'A review of demographic trends', p. 126, fn.13.
29. Tinker, *The Banyan Tree*, p. 44.
30. Tinker, ibid., pp. 20–21.
31. *The Economist*, 18 January 1992, pp. 60–81.
32. Tinker, *The Banyan Tree*, p. 142.
33. Tinker, ibid., p. 146.

34. Tinker, ibid., p. 150.
35. Tinker, ibid., p. 150.
36. Tinker, ibid., p. 154.
37. Tinker, ibid., p. 159.
38. M. Twaddle, 'East African Asians through a hundred years', in Clarke, Peach and Vertovec (eds), *South Asians Overseas*, p. 150.
39. Twaddle, 'East African Asians', p. 150.
40. Y. Ghai, 'Amin and the Asians', *Times Literary Supplement*, 4 July 1975, pp. 738–9.
41. Twaddle, 'East African Asians', pp. 155–7.
42. P. Bhachu, *Twice Migrants: East African Sikh Settlers in Britain* (London, 1985), p. 3.
43. B. Dahya, 'The nature of Pakistani Ethnicity in Industrial Cities in Britain' in Abner Cohen (ed.) *Urban Ethnicity* (London: Tavistock, 1974), pp. 77–118; M. Anwar, *The Myth of Return: Pakistanis in Britain* (London: Heinemann, 1979); V. Robinson, *Transients, Settlers and Refugees* (Oxford: Clarendon Press, 1986)
44. Bhachu, *Twice Migrants*; Vertovec, 'Community and congregation in London Hindu temples: divergent trends', in *New Community* 18, 2 (1992); 'Caught in an ethnic quandary', in Ballard (ed.) *Desh Pardesh*.
45. van Lier, *Frontier Society*, p. 217.
46. Ibid., p. 218.
47. Ibid.
48. Ibid., p. 393.
49 G.J. Oostindie, 'Caribbean migration to the Netherlands: a journey to disappointment?' in M. Cross and H. Entzinger (eds) *Lost Illusions: Caribbean Minorities in Britain and the Netherlands* (London: Routledge, 1988), p. 63, table 3.1.
50. *United Nations Global Report on Human Settlements 1986* (Oxford: OUP, 1987), table 1.
51. Cross and Entzinger (eds), *Lost Illusions*, p. 10.
52. D.B. Barrett, *World Christian Encyclopaedia* (London: OUP, 1982), p. 642.
53. See S.M. Bhardwaj and N. Madhusudana Rao, 'Asian Indians in the United States: A geographical appraisal' in Clarke, Peach and Vertovec (eds), *South Asians Overseas*, pp. 197–217; A.W. Helweg, *Sikhs in England* (Delhi: OUP, 1986 2nd edn.) pp. 214–20.
54. N. Buchignani and D.M. Indra with Ram Srivasta, *Continuous Journey: A Social History of South Asians in Canada* (Toronto: Ministry of Supply and Services, 1985).
55. A.W. Helweg, 'Indians in Australia: Theory and methodology of the new immigration' in S. Vertovec (ed.) *Aspects of the South Asian Diaspora* (Oxford University Papers on India, vol. 2, Part 2, Delhi: OUP 1991), pp. 7–35.
56. C. Peach, V. Robinson, J. Maxted and J. Chance, 'Immigration and Ethnicity', in A.H. Halsey (ed.), *British Social Trends since 1900* (London: Macmillan, 1988), pp. 567–70.
57. See C. Peach, *The Caribbean in Europe: Contrasting Pattern of Migration and Settlement in Britain, France and the Netherlands* (University of

Warwick Centre for Research in Ethnic Relations, Research Paper in Ethnic Relations, no. 15, 1991).

58. Peach, *The Caribbean in Europe*, p. 8.

59. See V. Robinson. 'Correlates of Asian migration to Britain', *New Community* 8 (1980), pp. 115–23; Robinson, *Transients, Settlers and Refugees*, pp. 27–8; Peach et al. in Halsey (ed.), *British Social Trends since 1900*, p. 590.

60. *Labour Force Survey* (1989), p. 29.

61 Peach, 'Estimating the growth of the Bangladeshi population of Great Britain', *New Community*, 16, 4 (1990), p. 488.

62. *Labour Force Survey* (1991), p. 28.

63. *Labour Force Survey* (1987), table 5.11.

64. *Labour Force Survey* (1991), p. 28.

65. *Labour Force Survey* (1989), p. 28.

66. C. Brown, *Black and White Britain: The Third PSI Survey* (Policy Studies Institute, Aldershot: Gower, 1984).

67. K. Knott, 'Bound to Change?: The religions of South Asians in Britain', in S. Vertovec (ed.) *Aspects of the South Asian Diaspora*, p. 91.

68. See B. Knerr, 'South Asian countries as competitors on the world labour market', in Clarke, Peach and Vertovec (eds), *South Asians Overseas*, pp. 173–96.

70. Knerr, 'South Asian countries', p. 178.

71. G. Battistella, 'Gulf Returnees: Profile of a Crisis', *Asian Migrant*, 4, 1 (1991), p. 25.

71. Battistella, 'Gulf Returnees', p. 26.

72. Tinker, *The Banyan Tree*, pp. 141–2.

3 The Modern Zoroastrian Diaspora

John R. Hinnells

The documented history of the (Parsi) Zoroastrian diaspora across India, to East Asia, Africa and the 'West' (mostly to Britain, America and Canada) stretches back to the mid-eighteenth century. It therefore provides a longer record of the 'trajectory' of the pattern of settlement of an Asian community than is available for any other group from the subcontinent. This chapter seeks to provide the first historical overview of this global dispersion, to identify the characteristic features of the various communities and to reflect on the impact of international migration on the preservation of traditions. Among the most important sources [1] of evidence are two recent world-wide questionnaires conducted among Parsis about their practices and attitudes.

A BRIEF HISTORY OF THE ZOROASTRIAN DIASPORA

The Zoroastrian diaspora began in the sixth century BCE when the Achaemenids established a large empire stretching from North India to Anatolia, Israel and Egypt. Officials and tradesmen travelled throughout the empire and beyond. Another major turning point was the escape into exile of Zoroastrians, oppressed under Islamic rule in their Iranian homeland, who settled in north-west India in the tenth century CE, thus establishing the Parsi (or Persian) community. That community settled first in Gujarat and from the seventeenth century migrated in increasing numbers to Bombay.[2] This chapter is concerned with subsequent travels.

The first record of an overseas sojourn related to Naoroji Rustomji who travelled alone to Britain in 1723, to protest to the Court of Directors of the East India Company against unjust treatment by Company officials in Bombay. The Court found in his favour and he returned to Bombay a wealthier man – and an influential one, for he used his 18 months in Britain to develop contacts.[3] His visit was important because it suggested to Bombay Parsis that there were considerable benefits to be gained from visits to Britain. He came and returned alone (apart from servants to prepare his food according to the purity laws) and so there is no suggestion of a 'community'.

The first Parsi settlement abroad was in the China Seas. Hirji Jivanji Readymoney migrated in 1756 but others came and stayed in small 'groups' from the 1770s (Readymoneys, Edulkakas and Camas). The oldest grave is in Macau, dated 1829. From the 1830s and 1840s one can begin to refer to 'communities': the Canton Zoroastrian Association was started on 19 September 1845; the Shanghai Association on 18 March 1854 and a building was opened in Hong Kong in 1861. Parsis came mostly to import opium and export tea, silks and spices. They moved into real estate and banking. The communities in China were expelled after the Communist revolution: some returned to India, others joined the Hong Kong community. They had already established a reputation there for charity, for example by starting Hong Kong university (Sir Homi Mody), a TB sanatorium and a rest home for sailors. As 1997 approaches the Parsi community has built a new tower block, including a community hall to replace the old one, as an act of faith in the future. In the early 1990s there are approximately 150 Parsis in the colony, mostly in business but one or two professionals (doctors, lawyers). Although the community makes regular charitable donations to local groups there is little interaction either with the Indian or the Chinese societies (apart from the social contacts arising through business). The ties back to India are reasonably close. The prayer room and funeral ground are maintained by a full-time priest and with regular weekly dinners the community is tightly knit.[4]

The 1830s and 1840s were decades of considerable Parsi overseas trade, travel and settlement in Sind, Aden, East Africa and Britain. They began to trade in Sind from the 1820s, but numbers and wealth grew after the Afghan Wars in 1839 when various Parsis (for example the Dinshaw, Minwalla and Ghadially families) travelled as brokers, agents and suppliers to the British. They moved into real estate and were noted for charitable work, particularly in the fields of education and medicine. The community there has built two temples, two schools, two Towers of Silence (*daxmas*), housing colonies, hospitals and dispensaries, a social centre and a library – despite the fact that at their numerical 'peak' (1911–40) they numbered only 4000–5000. Partition and the establishment of the Muslim state of Pakistan raised the concerns of some because of their historical memories of persecution. Responding to that fear, but even more to the attractions of education and career development in the West, a number migrated in the 1960s and 1970s so that in the 1990s there are only approximately 3000 left. The low birthrate, delayed marriage and the typical longevity of the Parsis in Pakistan mean that it is an ageing and numerically declining community. Despite the fears following 1947, Parsis have thus far flourished in Pakistan. Jinnah's wife, mentor and doctor were all Parsis. The early cabinet meetings were held in a Parsi hotel, the owner of which (Dinshah Avarai) went on to own the Lahore Hilton hotel and its equivalent in Karachi. His son became an MP in Islamabad, and several Parsis have held high government office: in the 1940s a Parsi was Mayor of Karachi (Mehta); in the 1980s a Parsi (Marker) was ambassador in Paris, Washington and then at the United Nations. Parsis have also held one of the highest law posts in the country and senior military positions. In broad terms the community is affluent, tightly knit and being conscious of the high profile of Islam, they perceive a religious gulf between themselves and the majority population. As mosque attendance is common, so also temple attendance among Parsis is high, not least among the young, for being religious is not considered socially unacceptable among peers. Intermarriage is rare, for marriage to a Muslim would inevitably mean conversion to Islam. Some Pakistan Parsis fear for the future if Pakistan becomes militantly Islamic, particularly in the smaller communities, for example Lahore, rather than in cosmopolitan Karachi where Parsis exist in greater numbers and have the material resources and influence to support each other.[5]

Aden developed as a port in the mid-nineteenth century and became important as a coaling station for steamers passing through the (new) Suez canal en route for the Indian Ocean and beyond. The pioneering Parsi family was the Dinshahs, or as they became known, the Adenwallas. Cowasji (arrived 1845) brought technical innovation to the port: a condenser to convert sea water into drinking water, an ice machine and a floating dock. An agent for British companies, Cowasji also began his own shipping line dealing with East Africa. His son, Hormusjee, developed further the family shipping business and diplomatic standing before retiring in 1920, aged 63, to Bombay, where he became influential in Parsi community and national politics. (He was a member of the Western Indian Liberal Federation, a conservative body with some Tory links.) In Aden the Dinshahs undertook considerable charitable work (educational, religious and medical, both communal and inter-communal) and they dominated the Parsi community, which grew to about 250 persons. In 1884 the Dinshahs expanded into East Africa which became a more important Parsi centre, although a dwindling community remained in Aden until 1976 when the sacred temple fire was moved to Bombay. In this context the significance of the Aden group is twofold: it exemplifies the business expansion of the community's interests and illustrates their technical innovation and their charitable activities; and it became an important stepping-stone in Parsi history for the community's move to East Africa.[6]

Whereas most Parsis went to Aden only for their working life and retired to Bombay, in East Africa (especially in the twentieth century) more people settled and reared their families. 'Africanization' in the 1960s drove most Parsis from the region onwards to Britain (and a few to Canada). These 'twice-migrants' became an important element in other communities and therefore merit more attention than the Aden community. Indian traders had a tradition of trade in East Africa and became a crucial element in Zanzibar, which came under increasing British control in the nineteenth century and was 'administered' mainly through Bombay. This connection, plus the time spent in exile in Bombay by Barghash before he came to power in 1870, resulted in a number of Parsis moving to the 'emerald isle' of Zanzibar. Some were professionals, others came as engineers, but most came as businessmen. When the Cowasjis came they built a fire temple – a full-time priest had already been funded – and with a funeral ground the Association (which had been formed in 1875) had the physical resources to maintain a religious life. The Zanzibar community reached its numerical peak after the Second World War but diminished with the revolution of the 1960s. In the 1990s only two Parsi families remained.

From 1870 Parsis began to settle on the mainland of East Africa, first in Mombasa and then at the turn of the century with the opening up of the East African railways they moved to the new centre of Nairobi. The two earliest leading figures were builders and architects, responsible for ports, law courts, treasury buildings, customs house, post office and banks. Lawyers, surveyors, engineers and doctors were the main professions pursued by Parsis in the early years in Kenya; more were businessmen and suppliers to the British. There is no evidence of any manual labourers among these Parsi settlers. Initially men came alone and returned to their families in India at the end of their employment, although occasional visits back to India were possible. An Association was formed in Mombasa from approximately 1897 and in Nairobi from about 1902 (when burial land was purchased). No temple was built on the mainland, although there was a community hall, with prayer room, and accommodation in Mombasa (which also had a priest from 1909), but Nairobi has not developed the same resources. The two centres have shared the services of the priest, but have not interacted a great deal otherwise. Both reached their numerical peak (c. 300 persons in each) after the Second World War, but diminished in the 1960s and at the start of the 1990s there are about 50 Zoroastrians in each city.

In none of these three East African centres have Parsis achieved the political and diplomatic status they did in Sind or Aden. Parsis in East Africa have remained a distinct community, mixing little with 'whites' or Indians and hardly at all with the majority 'black' population. Although they have not been able to carry out the full liturgical practices, lacking a fully consecrated temple and team of priests, nevertheless they have, as a distinct community, been reasonably free of 'acculturating' influences and can in some senses be seen as islands of nineteenth century attitudes and practice. When the Zanzibar revolution and the Africanization programme in the 1960s caused most to migrate they took with them this traditional mental framework.[7]

The history of the Zoroastrian community in Britain is more fully documented than that of other communities (Zoroastrian or South Asian).[8] It has a number of distinctive features. From the middle of the nineteenth century many came for education. Some travellers were businessmen who wanted to see the fruits of the industrial revolution at first hand in London, in the textile industry in Manchester and the port of Liverpool. Others were students, mostly coming for Law (in London) and Medicine (in both Edinburgh and London). Some also came simply for business – the first Indian firm established in Britain was set up in London and Liverpool in 1855. The number of visitors was such that a formal Association was started in 1861 and by the end of the century families were settling at 'the heart of the empire' as Parsis were fond of describing London. The perceptions of these nineteenth century visitors recorded in a number of diaries and travelogues show they were impressed by the industry and 'busy-ness' of Britain, its technological advancement and the 'splendours' of high society, above all of royalty. But they did not like all they saw, for example severe poverty in inner-city Britain, the divisions between the Christian churches, and the racial prejudice they encountered – specifically, in their opinion, from the 'lower orders' (the term used by the Wadia brothers in 1842). Parsis in India and in Britain were convinced that they were an important factor in establishing British power in western India, in the early military defence of the island of Bombay, as middlemen in trade, as negotiators and mediators on behalf of the British (e.g. with Muslim rulers), in building the Bombay dockyard and developing the port's overseas trade, in bringing the industrial revolution to India, and in pioneering various social reform movements. Parsis arriving in Britain therefore considered they had a good working relationship with the British by whom they believed they were respected.[9]

The declared aims of the Association founded in 1861 were to care for indigent Zoroastrians in Europe, to obtain and administer a funeral ground, obtained in 1862, and to support education, especially the study of Zoroastrianism. For the first 50 years of its existence, under the presidency of Dadabhoy Naoroji, the Association was essentially an informal burial club, but under its next president, Sir Muncherji Bhownagree, it became a formal incorporated body owning its own property in London which functioned mainly as a social centre but also had accommodation for visitors to stay, and a prayer room. Until the Second World War numbers were approximately 150–200 but thereafter grew reaching approximately seven to ten thousand in the 1990s. Numbers increased for various reasons: several medical people migrated to Britain when the National Health Service was started in 1947; in the 1960s Parsis migrated from India and Pakistan along with other South Asian groups for education and a career when fears grew that Britain was going to restrict immigration; also in the 1960s there was an influx of Parsis from East Africa when South Asians were forced to leave; in the 1970s a number of Iranian Zoroastrians came to Britain, again mainly for education (some for careers), and a number settled after the fall of the Shah and the establishment of the Islamic Republic in 1979.

In comparison with other Parsi diaspora groups British Zoroastrians have a high profile within national politics. The first three Indian MPs were all Parsis: Dadabhoy Naoroji, 1892–5, Liberal; M.M. Bhownagree, 1895–1905, Tory, and S. Saklatvala, 1922–8, who was first elected as a Labour MP and then as a Communist. All were elected for London constituencies. They were very different personalities, as well as in their political allegiance.[10] Naoroji saw himself as the MP for India and sought election in order to present the case regarding the 'Condition of India' that the Indian National Congress was, with increasing frustration, trying to present in India.[11] The themes which he pressed were that Britain drained away substantially more in revenue (e.g. through taxes) than it invested; it perpetrated a gross injustice by effectively excluding Indians from entering the Indian Civil Service; and finally robbed them of a role in administration of their own country.[12] Bhownagree is commonly represented as a Tory counterpart to Naoroji and uncritical in his obsequious acceptance of the British. Although he was deferential, at the heart of his message, especially during his second term in parliament, was a serious questioning of British economic justice in its dealings with India. He called for a greater emphasis on technical education so that India could become more economically independent. He was also at the forefront of the campaign on behalf of the Indians in South Africa and a mentor of Gandhi's work there. Saklatvala's primary concern was to see justice for the working classes who he was convinced were unjustly oppressed by wealthy rulers the world over. He condemned Britain as the most bloodthirsty nation in history, for the British, he said, not only killed those who challenged them, but also defenceless peoples in their own homes in Africa and Asia.

It is difficult to assess the impact of these MPs. Naoroji was successful in campaigning for a Royal Commission on the Indian budget, but the

Government effectively determined the outcome by the members it appointed. One way in which they were important was in the pride and self-image their elections effected among the Parsi community. Why were these three successful when other attempts by Indians (before the 1980s) to obtain election all failed? Evidently each was supported by their party – but why did the parties choose them? Naoroji was an educated man (at 27 he was the first Indian to be appointed a professor). His appearance and manners made him seem British, despite some opponents describing him as a 'blackman' (the Prime Minister, Lord Salisbury), and the 'right wing' press accusing him, on his election, of being a fire-worshipper. Bhownagree was a (constitutional) lawyer, very western in appearance and manners and a good speaker. Saklatvala was less well educated but came from India's leading industrial family (the Tatas) and so was accustomed to moving in 'exalted' society – however much he came to reject it as a Communist. He was respected even by his opponents as not only an able speaker but also as a man of great courtesy and civility.[13] He was successful because of Labour support in a Labour stronghold but it is remarkable that neither Naoroji nor Bhownagree contested such 'safe seats' – their achievement is, therefore, all the more remarkable. The three of them reflected, in different ways, typical Zoroastrian virtues, above all an absolute commitment to justice, though how each of them saw the focus of that quest for justice was different. Saklatvala, for example, went to prison rather than agree to stop speaking on behalf of the striking workers at the time of the General Strike and could probably have been Under-Secretary of State for India had he been willing to follow the Labour Party line on India. Each of them benefited from being seen as members of a 'marginal' group whose success did not threaten the standing of other groups (as for example a Hindu or Muslim might have done). They were all prepared to work within the existing British political framework. However much they wanted to reform it, none of them supported violent means of protest. The Parsi role in British politics has not ceased, for there is at the time of writing another, Zerbanoo Gifford, who is an important figure in the Liberal Democrat Party, campaigning on the rights of ethnic minorities and especially on women's rights. In Britain, therefore, Parsis have seen themselves as continuing their involvement in the running of government as they had in nineteenth and twentieth century British India. This same willingness to identify themselves with the majority population, more than they have in Muslim Pakistan, Chinese Hong Kong or Black Africa, has perhaps left them more open to the forces of acculturation.

The first Parsi known to have visited America was J.H. Kothari in the course of his world tour in 1883/4. Three of the first Parsi settlers in the 'new world' became millionaires at the turn of the century: P.S. Saklatvala (in the oil business); N. Heeramaneck (art dealer) and P.F. Davar (business unknown).[14] As a result the American image as a land of opportunity was reinforced. Some elder American Zoroastrians remember that there were 12–20 Parsis studying in New York at any one time in the early decades of the twentieth century.[15] The most influential of these was M.N. Dhalla (1905–9) who studied for a PhD at Columbia University under the distinguished Iranist, A.V.W. Jackson. His studies included courses on Comparative Religion. He also had contacts with various Christians; above all his Protestant mentor, Jackson, exerted considerable influence on him. In his own words, Dhalla who had been a strict orthodox in Karachi eventually left the shores of America a staunch reformer. When he was elected high priest (Dastur) of Karachi, he was known as 'the Protestant dastur' because of the way his beliefs and practices were thought to be influenced by that tradition.[16] Dhalla wrestled with integrating his devotion to his religion with Western learning and assumptions in a way which not only prefigured spiritual wrestlings of Zoroastrians in the West at the end of the century but which many have found helpful so that his books are often quoted in newsletters and pamphlets produced by various associations in Britain and especially America.

Until the mid-1960s immigration from Asia to America and Canada was restricted to those with needed talents and was essentially restricted to Europeans. So it was estimated that there were less than 300 Zoroastrians in the two countries in 1965 when the Hart-Celler Act changed American immigration policy and allowed an equal number of migrants from Asia as from Europe. Canada changed in a similar fashion soon afterwards. The earliest Zoroastrian centres were Montreal, Vancouver and Chicago in 1965. Further centres quickly developed, notably in Toronto, New York, Los Angeles and Houston. Over the 25 years, Associations acquired their own buildings in five different cities, and formal Associations were established in a further 16 and an umbrella organization, the Federation of Zoroastrian Associations of North America (FEZANA), was started in 1987. There is a mixture of countries of origin in America and Canada though in comparison with Britain there are far fewer East African but many more Iranian Zoroastrians. The motivation behind migration was higher education, especially in the sciences. Ninety-two per cent of Zoroastrians in New York and Chicago have been to university, most (62 per cent of the total population) to pursue postgraduate studies. Others have moved to the continent because of employment with multinational companies. The American and Canadian Zoroastrian population is therefore mostly of highly educated professionals and relatively few business people. They display a high degree of organization and a strong emphasis on religious education as a means of enabling the young to resist the acculturation pressures of the 'melting pot'.

The most recent, and increasingly popular, target of Zoroastrian migration is Australia, especially Sydney and Melbourne. The first Zoroastrian I have been able to trace was Hirji Lentin who opened a textile business in the mid-1920s. The first Association was started in Sydney in 1971 but a building was not opened until 1986. The Melbourne Association was started in 1978 and its constitution was formally incorporated in 1987. By 1990 there were approximately 500 Zoroastrians in Sydney and just over 200 in Melbourne. Australia is proving increasingly popular with Bombay Parsi youth partly because of the perceived career opportunities, and the climate which is more attractive than the Canadian winters. Higher education is not as attractive a magnet on this continent.

The diaspora communities all have a different demographic profile from the 'parent' community in the old country. Data are not available from Iran, but in India and Pakistan the community is an ageing one, with a growing proportion over 60 years of age. There are fewer marriages, delayed marriages (as young people pursue education and careers) and fewer children in each family.[17] The result is that the Indian community decreased by 20 per cent in the decade 1971–81. The age structure is also different, for it is mostly the young who have migrated and relatively few have been able to take their parents and grandparents with them. Thus although diaspora Zoroastrians also have small families (an average of less than two children was indicated by the Survey) overall numbers are growing because the young average age of the community means that the death-rate is lower than the birth-rate and some migration continues. The diaspora is at once 'draining away' numbers and many of the potential leaders because it is the highly educated and the 'achievers' who are settling overseas.

DIASPORA ZOROASTRIANS: THEIR SETTLEMENT AND RELIGION

How different is it, being a Zoroastrian in, say London, Hong Kong, Mombasa or Chicago? What follows will have to begin with generalizations which will in turn require qualification. Many of the terms used, for example 'acculturation', require discussion. 'Acculturation' does not indicate a clearly identifiable or consistent process: not only do different individuals adopt the majority culture in different ways from each other, some individuals adopt the culture in one aspect of their lives, such as language, but not in another, such as marriage customs. One 'shorthand' term used in this chapter is 'traditional' as it is less emotive than the labels commonly deployed within the community, namely 'orthodox' and 'reform'. The criteria by which respondents' level of 'traditionalism' was measured were frequency and especially manner of prayer (for example language used); religious practices (for example lighting a *divo* or 'oil lamp' in the home and festival observance); attendance at a religious centre (where practicable); attitudes to the purity laws and intermarriage and links with the old country through correspondence and visits.

ORIGINS AND MOTIVES

Where people come from affects how they settle after migration. Typically those who have migrated from rural Gujarat preserve the Zoroastrian traditions more fully than those who have migrated from cosmopolitan Bombay. So more people from Gujarat tend to continue to think in Gujarati, compared with people from Bombay who generally think in English; they pray more frequently and more traditionally (i.e. in Avestan, not the vernacular) and they hold more 'traditional' views in opposing intermarriage. They also tend to keep in closer contact with the 'old country'. The explanation is that Parsis living in westernized Bombay have already begun the process of adaptation which people moving from Gujarat have not taken so far. Other trends are also clear. Zoroastrians who have migrated from Pakistan are more traditional than the Bombay Parsis. In part this would appear to be due to the Religious Education Programme which is stronger among Karachi Zoroastrians than in Bombay. A major influence in this was undoubtedly Dastur Dhalla who inspired the teaching in the Karachi Parsi schools. His policies have been continued by his pupils so that they have affected recent as well as earlier migrants from Karachi, and they have travelled with a clearer perception of their religious and cultural heritage than have many from Bombay. Another factor in the Karachi tradition is that living in a Muslim environment Zoroastrians typically see it as potentially hostile and therefore remain behind their own 'cultural walls' and acculturate relatively rarely. Through Dhalla, however, they have been subject to the 'protestantizing' influences of Western education.

Those from East Africa have travelled with a different set of 'cultural luggage'. As noted above, they have typically preserved more of the community's early traditions. They have been relatively isolated not only from the culture of the majority 'black' population, but also from the

British in East Africa, who were more distanced from the Indian community than they were in contemporary Bombay. Zoroastrians in East Africa have, therefore, not been subject to as much influence from the evolving Westernizing influences of their educated co-religionists in the commercial capital of British India. When the East African Zoroastrians came to Britain their existing set of networks were stronger than any with the existing Zoroastrian community in Britain, and indeed there are several references in the London records of the Committee expressing concern that they were not proving successful in establishing contacts with the new arrivals.[18] In the 1980s there were serious divisions within the London community due to a complex web of factors, including clashing personalities, and the belief of the new committee that their predecessors had been too 'liberal' in their adaptations of traditions (for example over non-Zoroastrians at death ceremonies). It would be simplistic to say that the new 'traditional' committee is identified with East African Zoroastrians and the 'old' with Bombay and Karachi, yet there is an element of that division.

Zoroastrians are different from other South Asian communities because there is an important group from outside the subcontinent: the Iranian Zoroastrians. There have been difficulties in integrating the co-religionists, so much so that in New York, Los Angeles and San Francisco separate groups have been formed, and formal divisions have been feared elsewhere. When the two groups viewed each other across the Hindu Kush the Indian and Iranian communities were able to stress their identity, but living together after migration has proved difficult. At a superficial level there have been disagreements over food at communal festivals: Parsis do not enjoy kebabs and Iranians found Parsi food too spicy. Language barriers have proved a problem as have the different calendars of festivals. But the differences lie deeper also. Living in a Muslim environment Iranian Zoroastrians see all authority as lying exclusively in the words of the Prophet as revealed in sacred scripture; they emphasize the abstract nature of religion and often speak negatively about the role of a priesthood and of rituals. Parsis from India have existed in a Hindu environment and commonly see the priest as a man of spiritual power and complex rituals as points of access to that power.[19] There is also a dispute about temporal power: traditional (though by no means all) diaspora Parsis accept the authority of priests and leaders in Bombay but not that of the priests in Iran, whereas Iranians in the diaspora would not accept the Indian authorities at all. Thus, in terms both of spiritual and temporal authority and perceptions of religious experience there is a marked difference between the two groups which it has so far proved difficult to reconcile.

In short, the place whence Zoroastrians migrate affects how they settle in the new country. In addition, the motive behind the migration affects the patterns of settlement. Where people have migrated for their own self-perceived benefit (such as for education or career as Bombay Parsis did) then they tend to accept and adapt more to the new 'home' than do those who have felt forced to migrate, as did the East African and in particular the Iranian Zoroastrians. The last group have a stronger 'myth of return' to the homeland, feel closer emotional ties to it (even though questions of safety mean they cannot yet easily go for visits) and maintain the language and allegiance more strongly. Many Iranian Zoroastrians strongly assert that they are the true, original, Iranians: their ties to the 'homeland' are therefore especially strong.

Different patterns of settlement, especially in religious matters, tend to correspond to the types of people who migrate. The essentially business communities (Karachi, East Africa and Hong Kong) are more traditional than those with a high proportion of educated professionals (e.g. America and Canada). Businessmen who travelled overseas in the nineteenth century were innovative individuals and their activities required them to take what was then a 'liberal' attitude to certain traditions, such as long sea voyages which involved defiling the sacred element of the waters. But it is remarkable how many of them sought to preserve the purity laws, for example by taking servants with them rather than inter-dining. Further, those who were successful had the funds with which to erect buildings, even fire temples, and employ a full-time priest (in the China Seas, Sind, Aden, East Africa), whereas the professionals have rarely acquired such fortunes, and so communities they dominated (e.g. in Britain and many centres in America and Canada) have had fewer material religious resources. The assertion needs some qualification, because some wealthy businessmen have funded religious resources (notably the Iranian Rustom Guiv who has funded most of the centres used and largely dominated by the professionals in America), but broadly the two groups have different values and resources. Further, highly educated individuals commonly seek to articulate and reason their religion in order to elucidate both for themselves, their children and their non-Zoroastrian friends the nature of Zoroastrianism, in a way which typically business people do not, for they commonly see their religion as part of their life to be practised, not reasoned. Few communities consist entirely of one section of society and there are exceptional businessmen and professionals, but the exceptions do not invalidate the broad picture of a business community tending more towards a traditional religious pattern than do highly educated professional classes.

THE DESTINATION OF MIGRATION AND ITS IMPACT

New migrants commonly have as many stereotypes concerning the majority population as the wider society does of them. Generally interviewees indicated that they viewed the British as cold, remote, unfriendly, untrustworthy, sexually promiscuous and lacking in a sense of family responsibility. Britain was broadly seen as a very secular, and therefore religiously threatening, society. In America, Canada and Australia such negative images are less evident, though some do emerge, of America above all as a 'melting pot' and as a place where religion is a highly publicized dimension of life both in missions on the street and television and the influence of evangelicals and the 'moral majority' on government. The major threat is seen to be the dilution of individual cultural identities in the quest to produce the 'real American'. Canada is, according to our recent Survey (see note 1), seen as less of a threat. Although the country is known to have had a strong racist element in recent history it is typically seen as having a powerful multicultural programme which encourages individuals and groups to preserve their identity (so, for example, the Association in Toronto received a substantial grant to enable them to extend their building, without any conditions regarding access for non-Zoroastrians). It is probably not coincidental that more Zoroastrians in Canada were willing to identify themselves as Canadians than were their co-religionists in the United States willing to identify themselves as Americans.

The actual experiences in a new society inevitably affect both individuals and groups. In the Survey respondents were asked if they considered they had met racial discrimination, and if so, whether it was in specific dimensions of life (education, employment, housing, police, or in public places) and whether it was an experience they considered they frequently/sometimes/rarely or never had. The city with the highest proportion of individuals considering that they faced discrimination was Toronto (48 per cent), but most of them said that it was rare (usually in initial moves to obtain employment). Forty-two per cent in London said they considered they had experienced prejudice but in contrast to Toronto most of them said they thought they experienced it more often and more substantially, mostly at school, than in career terms and much less in housing and little from the police. These figures, obviously, do not measure actual prejudice. They probably underestimate it, for several people said they did not face prejudice, but had experienced 'Paki-bashing'. Many implied that racial prejudice was something 'others' experienced, because it was assumed that they themselves 'were above that sort of thing because you see I am Westernized'. But the data indicate that contrary to popular 'myth' there is no one type of person who provokes prejudice. Those who considered they had experienced it were as much those who thought in English as in a non-English language, the educated, the professional high flyers (perhaps the group which most experienced the problem were people in the medical profession), and those who were born here more than those who migrated in the 1960s. Prejudice would, therefore, seem to be in the eye of the beholder rather than in any type of 'victim'. The impact of that perceived prejudice depends on a variety of factors. In various ways Zoroastrians in Hong Kong and in East Africa would consider that the majority Chinese and 'blacks' respectively are prejudiced against South Asians, but they appear to feel less affected by this than in Britain, America, Australia and Canada, where their lifestyles and desired social acceptability are at stake. The impact of prejudice seems to be to provoke individuals to assert their identity rather than to acculturate to avoid prejudice.

The physical location, as well as personal experiences, inevitably affect patterns of settlement. There is a different pattern of practices and attitudes between Zoroastrians living in a 'centre' (for present purposes defined as having a formal Zoroastrian Association with a meeting place) and those living in the 'periphery', that is, without such a Zoroastrian support system. In Britain, for example, there are different levels of practice between Zoroastrians in London and Manchester. Because the former can meet regularly at Zoroastrian House their level of community contacts is greater, their wearing of the sacred shirt and cord (*sudre* and *kusti*) is higher and so is the saying of the associated prayers; and people tend to keep in closer contact with the 'old country'. Since the Survey was conducted, a Zoroastrian Association has been started in Manchester which seems to have led, among some individuals and their families, to an increased level of Zoroastrian activity. Certainly the Survey in America and Canada indicates that where individuals are deployed by their companies to work in areas distant from fellow Zoroastrians then, because it is difficult to maintain the practice, the ties decrease emotionally as well as physically. Undoubtedly the prejudice and persecution associated with ghettos are undesirable, but ghettos have the merit of facilitating communal contact and support, thus reducing pressures of acculturation.

Each community has its own distinctive history which can produce different Association 'characters' even within the same country. A striking example for Zoroastrians are the cities of Chicago and Houston. In the Survey the responses from these communities displayed a contrasting pattern: broadly, Houston appeared to be more 'traditional' than Chicago and several reasons for this emerged on further study. First, Chicago is one of the older American associations and therefore has a high proportion of 1960s migrants who, typically, tended to be the more 'liberal', innovative and Westernized sections of educated Parsi society. Houston, in contrast, is relatively 'new' in that most of the members migrated there in the post-1980 oil boom and generally migrants tend to be more traditional for the first 10–15 years of their settlement. Secondly, Chicago has attracted a very high proportion of educational and professional 'high flyers' and scientists. In Houston, however, the Zoroastrians are mostly businessmen attracted there by the booming economy. Thirdly, whereas Chicago is a predominantly Indian (mostly Bombay) Zoroastrian population, in Houston there is a substantial section of the community from 'traditional' Karachi. 'Chain migration' is a major factor in this pattern, for a Karachi family was among the early arrivals and they encouraged friends and relatives to come also. Fourthly, within small diaspora communities individuals can play a very significant role in shaping that community (an important, if somewhat obvious, point to make since surveys identifying trends can falsely imply that when certain factors combine, such as age, gender, education, then the pattern of settlement is determined). Within the Chicago community prominent figures, both priestly and lay, have emphasized the need to adapt to the American continent, whereas Houston has some quite traditional leaders. Finally, there is a different history of migration to the two cities. In Chicago most Zoroastrians arrived as students and only gradually were they able to marry and set up a family and relatively few achieved sufficient wealth quickly enough to bring their extended family. In Chicago, therefore, there is a higher proportion of nuclear families. In Houston, in comparison, most came as reasonably successful businessmen and went straight into well-paid jobs and were therefore able to bring their families with them – including grandparents.[20] Where grandparents are present, especially grandmother, they appear to act as particularly powerful forces for the preservation of tradition. The important point suggested here is that it is wrong to refer to 'American Zoroastrianism',[21] as though it were a monolithic phenomenon. In an international perspective, there is even greater diversity within 'diaspora Zoroastrianism'.

SOME COMMON 'TRAJECTORIES'

Despite the diversity of 'Zoroastrianisms' in the diaspora it is possible to identify some common trajectories. From the international Survey it would seem clear that, typically, the Zoroastrians who migrated to Britain, America and Canada in the 1960s (and similarly those who migrated to Australia in the 1970s and 1980s) were the Westernized, innovative people most likely to acculturate fairly quickly. Those in the vanguard of the migration are commonly the most 'adaptable' people. Those who have come later tend to be traditional. However, once an individual has been settled in 'the new world' for 10–15 years the process of adaptation often becomes clearer, whereas in the early days of settlement not a few try to stand firm against what they then see as the erosion of their identity.

Many individuals, particularly those born in the 'new world' follow a similar pattern of development: they follow the religion of their parents until late teens when they leave home for work or (more often with Zoroastrians) to go to university. At that point many turn their back on their tradition, but as marriage approaches (especially if it is marriage within the community) then they begin to build bridges to their heritage, bridges which in a number of cases become highways when they have children of their own. The Survey would suggest that there is a strong connection between having a family and seeking to preserve the heritage.

But other factors may affect these common 'trajectories'. In terms of gender differences, women tend to be slightly more religious than men (frequency and manner of prayer, attendance at a centre of worship) but markedly stricter on issues relating to the family, for example, on issues concerning race and religion, especially intermarriage, and observance of the purity laws. There can be counter-factors: if men do go to the centres of worship they typically tend to go more often and become involved in committee work. But where women go out to work the pattern of their religion approximates much more closely to that of the men, in other words, it is less a gender difference than the role of the housewife.

Education is, inevitably, a major factor in the preservation of heritage. Those who had a primary education in the 'old country' tend to practise their religion more than do those who had a primary education in the West. The location of secondary schooling and the first degree does not appear to be as significant, but whether someone goes on to a higher degree and research does make a significant difference. Those who undertake post-graduate studies in the sciences typically affirm what might be called their 'ethnic links' (food, language, music); more strongly but substantially fewer of them assert their religious links. If such a survey were conducted among any community in Britain, America or Australia the same conclusions might be drawn. The problem of Zoroastrians is that such a high proportion of their diminishing numbers pursue these studies that the long-term threat such education poses for them is correspondingly greater.

There is a popular image of young Asians as distanced from their elders and rebelling against parental pressures to conform to community traditions, and as distancing themselves over two or three generations from their religion as they become progressively Westernized. The Survey suggests that this is far too simplistic a picture. Certainly the global responses suggest that the young pray less often, pray less traditionally, observe fewer religious practices, attend Zoroastrian centres less often and do not keep in such regular contact with fellow Zoroastrians. But were the present generation of elders any different in their youth? Is religious activity commonly something which increases with age? Another way of looking at the figures is that it is remarkable how many young Zoroastrians said they prayed regularly (approximately 40 per cent globally) – more than their non-Zoroastrian contemporaries, if the answers are an accurate reflection of actual practice. But this broad statement, like so many, requires qualification. What also emerged was that when the young people do engage in their religion they commonly do so with particular vigour and on many issues can be stricter than their elders. For example, a substantial proportion of the youth seem to be engaged in a quest for their roots and among this section there is a strong desire to preserve community boundaries, for example, in terms of in-marriage and of not allowing non-Zoroastrians into prayer rooms. Thus one 16-year-old respondent in Toronto commented that every day of her school life she was in contact with non-Zoroastrians and she wanted one place where she could go and be herself. Many in each country commented on the need to keep explaining oneself when not in Zoroastrian company. Interestingly, when asked about visits back to the 'old country' a higher proportion of the youth said that they had found such visits stimulating than did those who had migrated to the 'new world' in the 1960s – again an indicator that this particular section of the older generation appears to be one which has acculturated more than others, more than the young people born in the West who seek to affirm their identity. Thus it is a mistake to assert simply that the younger generation is less religious. The extent to which they are involved in the religion is dependent to a significant extent on the level of activity the elders allow; hence some centres have far more active youth programmes than others, for example by having youth representatives on the managing committee and allowing youth to organise their own social activities. Many young Zoroastrians are alienated by the feuding which characterises so many groups. From an external perspective it is evident that the communities need to unite in the face of the threat posed by the demographic decline and dangers of dispersion. But precisely because so many are acutely aware of the danger of extinction views held about future policy are held with particular vigour and the disagreements can be very vitriolic because each fears the 'others' are threatening the survival of the whole community. Given the strong family ties, it is very common for whole families to become alienated as their elders dispute with each other. Young Parsis, therefore, often feel caught between at least four factors: the acculturating pressure of their peer group at school/college or workplace, their quest for their roots, the internal community debates, and the great difficulties of practising the religion in the diaspora where there are simply not the physical provisions for the religion (for example, pure temples and a team of priests).

The transmission of tradition from generation to generation is a serious problem, especially for parents educated in a different continent and by different methods, such as for those brought up in Parsi colonies in Bombay where religion is something 'caught not taught', a matter of community membership not of rational elucidation. Such a background does not equip them to help their young brought up in a Western educational background, not least in Britain where there is a formal Religious Education programme in schools which is in practice Christian-based and commonly unavoidable (it is, for example, effectively impossible to opt out of the Christmas play which dominates school life for days, if not weeks, in junior schools).

Intermarriage is typically seen within the community as one of the greatest problems it has to tackle. The tradition is that in cases of intermarriage the offspring of a Zoroastrian male may be initiated (following a 1906 High Court test case in Bombay) but that a Zoroastrian woman's children by a non-Zoroastrian may not be. Commonly Zoroastrians assert that there should not be such gender differences in the religion; therefore in Bombay the 'orthodox' priests argue that no children of any intermarriage should be initiated, but in the diaspora there are growing calls to accept the children of any intermarriage. The issue has, for example, become an extremely contentious one in 1992–3 in London where the World Zoroastrian Organization (partly because of its consciousness of trends in other countries) has decided to accept partners and offspring as full members and has been vigorously attacked for doing so by what has become a more 'traditional' Zoroastrian Trust Fund of Europe, also based in (and almost wholly concerned with) Britain. It is a matter of great debate in Australia, but here it seems that in this younger, more 'liberal' community it is a less acrimonious debate, despite differences between Sydney and Melbourne. The Survey suggests that the people who intermarry are most typically the very well-educated, especially the post-graduates in science, and the career 'high flyers'. Not surprisingly it is much more common in the smaller Associations and among people living in 'the periphery' than in the larger Associations where young people have more opportunity to meet potential partners. The Survey suggests that, typically, intermarried Zoroastrians pray less often, pray less traditionally, observe fewer Zoroastrian practices and keep in touch with co-religionists less frequently and attend centres of worship less often. There is no way of knowing which is cause and which is effect. Is it that the less religious tend to be those who marry out or that out-marriage tends to distance individuals from the community? Undoubtedly in some instances the community distances itself from the intermarrieds. Although intermarriage occurs, and is a matter of very public debate in the 'old country', not least in Bombay,[22] it is much more common in the diaspora and is naturally a phenomenon associated with migration.

RELIGION

Most studies of ethnic minority communities focus heavily on housing, education and discrimination and neglect religion. But religion is also a highly significant issue. Almost all the author's respondents, interviewees and personal contacts asserted strongly that they considered they had become more religious after migration than they were before. Contrary to a popular image migrants do not leave their religion behind in Asia. As noted above, for many of the young people born in the West religion is a crucial part of their identity which they wish to assert. The definition of 'religion' needs expanding. Approximately two-thirds of respondents said they considered religion was a matter both of belief and of race (less than 20 per cent thought it was simply a question of belief). For most Parsis religion is something that is 'in the blood' or 'a matter of genes'. Zoroastrians have a particular problem in asserting their identity. They want to assert that they are not simply 'Indian' or 'Pakistani' or 'British' or 'American'. They may identify with both the 'old' and the 'new' country to a substantial extent, yet they want to assert their distinctiveness. Many want to assert that they are the 'original Persians' (since 1979 the term 'Iranian' is less popular because it is thought to have connotations of Islamic fundamentalism). Many also, however, want to transcend national boundaries and to assert that they are simply 'Zoroastrian' – and this may not indicate a theological allegiance, for a number of interviewees expressed the view that one could be both a Zoroastrian and an atheist. Even if the Zoroastrian is a theist, there remain severe problems of identity in that few 'outsiders' would recognize what a Zoroastrian is. Public ignorance about the religion (or race) means that it is difficult for Zoroastrians to assert the identity that they wish to claim.

For a sense of identity to be strong it is necessary for it to be 'meaningful'. What was meaningful for an older generation living in a Parsi colony in Bombay may not be meaningful for an American or Australian High School student. It is difficult for the religion to be the same in the diaspora since there are not the material resources such as a pure temple where the traditional expressions of devotion and ritual can be conducted, nor are there full-time priests able to dedicate their lives to the pure pursuit of spiritual power as there are in the 'old country'. Funerals can also present a major problem. In India the traditional Zoroastrian funeral involves exposure of the dead to vultures in towers of silence and an associated rich temple tradition of rites and prayers for the dead whereby the soul is aided, grief is channelled, respect is shown and duty performed. This is simply not possible in the diaspora. My own observation, having lived with Zoroastrian families when bereavement has unexpectedly occurred, is that after a death in the 'new world' natural grief is compounded by a sense of guilt because mourners are not observing fully what they were taught from infancy to be their duties for a loved one.

Different needs and different resources therefore generate a distinctive religious expression in the diaspora. This is a subject which has hardly been studied.[23] Within the Zoroastrian community there have been lively, indeed bitter, debates on who can guide any such development. Many in the diaspora feel that only those living outside India can understand the needs and pressures in their situation. There have, therefore, been meetings of priests (*mobeds*) organized in Toronto to try and define a consistent liturgical practice for the continent. The need is acute for all the priests are part-time. In their youth they received the conventional training for boys in their early teens but few, if any, of them when they migrated for education or business thought that they would function as priests. It is only in the wealthy nineteenth-century business communities (Karachi, Aden, Zanzibar, Mombasa and Hong Kong) that priests have been recruited for sacral duties from India. None have been recruited from Iran. The diaspora priests would not, therefore, see themselves as 'professional' priests in the traditional sense. An interesting, though still very rare, development is for youth in the diaspora communities to spend sometime at a priestly college in Bombay, to return and minister to their co-religionists. Many of the traditional high priests (*dasturs*) in India strongly object to diaspora 'part-time' priests changing (which virtually always means shortening) the hallowed rites for the convenience, as they see it, of those who have migrated for their own personal benefit. Thus, on the one hand the diaspora communities consider the traditional authorities do not understand their needs and on the other the *dasturs* consider that diaspora reforms are done for personal convenience, not for religious reasons. Because there is no clear religious authority in Zoroastrianism, comparable to the Pope or the Dalai Lama, it is difficult for Zoroastrians to obtain a clear guide on what is and is not theologically acceptable.

Change nevertheless happens. In the Survey more Zoroastrians consider they have changed their practices than they have their beliefs. From an outside perspective it would seem likely that if rites have changed, then so have beliefs, even if the practitioner is unaware of the latter. As already indicated it is evident that practices have changed, for example the absence of a pure fire temple. My own observation suggests that a distinctive liturgical pattern is evolving. An age-old rite in Zoroastrianism is the *jashan*. It is celebrated by a number of priests in any 'clean' place – it can be the home or in a public place or a temple. By changing some of the prayers it can be used either in memory of the dead, as an act of thanksgiving, as a communal act of petition or to honour a newly appointed high priest.[24] Because this ritual is not dependent on a temple and is a very adaptable rite, it has become a common part of diaspora communal religious activity, a focal (or sometimes a starting) point of community gatherings. Traditionally, although many can and sometimes do attend, in practice such rites are generally private ceremonies as are almost all formal acts of worship in Zoroastrianism. This perspective on congregational religious practice harmonizes with the natural wish for co-religionists to meet and socialize. It is easy to overlook the wider importance of such apparently trivial social activities as bingo or discos, for they can provide rallying points whereby people come together to exchange not just news, but to reaffirm and reinforce their identity. They provide a chance for Zoroastrians to be themselves without having to explain or justify, to relive memories, to maintain social networks and not least to give the young opportunity to meet peers who share their background. Being the only Zoroastrian in workplace, school or college can leave some young people not only ignorant but feeling awkward about their identity. Linking such crucial activities with a religious function, such as a *jashan*, adds the specifically Zoroastrian dimension to the occasion. But many, not least the young, also feel the need to have the rite and its significance explained to them for in a Western environment there is a felt need to explain 'rationally' why everything is done as it is. Custom and tradition (rightly or wrongly) are not often accepted as acceptable justifications. The result is that elucidations of rites and beliefs are developing.[25] The most 'productive' area for this is America and Canada, partly in Sunday-School booklet form (Chicago was among the leaders in this) and newsletters. These tend to present Zoroastrianism in what may, perhaps, be summarized as a Protestant format, that is with a perception of prayer as a means of communication or dialogue with God so that understanding the meaning of the words by using the vernacular or having translations is assumed to be necessary if prayers are not to be seen as 'mumbo-jumbo', in contrast to a traditional Indian (Parsi) attitude that prayers involve making alive through recitation the spiritually powerful words which invoke and make present the divine forces.[26] Traditional myths are allegorized or ignored, evil is thought of as an abstract force within man, and the only authoritative texts are seen as the words of the Prophet himself and later priestly writings are seen as medieval 'ecclesiastical' corruptions. Rituals, especially the aspects associated with the purity laws, are simplified and expressed in terms of Western scientific concepts, such as hygiene. This theological interpretation was prefigured in the writings of some Parsis in India who had pursued Western approaches to the study of Zoroastrianism, such as Dhalla and Sir J.J. Modi for they too had tried to integrate their traditional faith with modern knowledge.[27] It is these rather than the 'orthodox' writers who most diaspora Zoroastrians turn to, although there are indications that some in the diaspora are beginning to turn to the traditional high priests and writers. If these indicate a developing trend, together with the impact of East African Zoroastrians and more recent arrivals, and with the 'traditionalism' noted among the young 'questing for their roots', then the popular image of migrant communities acculturating progressively until they disappear, is a false image of the trajectory of settlement. The Zoroastrian diaspora history would suggest that a reaction against what is seen as 'excessive' liberalism may well develop and seek to reassert what is seen as the 'true' tradition.

CONCLUSION

There are many ways in which the Parsi diaspora represents a macrocosm of the microcosm of South Asian migration to the West: they come from India, Pakistan and East Africa. They have a longer history of migration than other communities and much can be learned from the study of this history regarding the patterns or trajectory of settlement, but they travelled in their greatest numbers at the same time (from the mid-1960s) as other groups did. They therefore became subject to the same laws, restrictions and prejudices. In many ways their religious and social needs replicate those of others, for example the importance attached to in-marriage, the need for community networks for mutual support, the evolution of an 'appropriate' religious tradition. However different Zoroastrians consider they are from other South Asian groups, they are seen by most 'outsiders' as a part of an imagined 'whole' of Indians/Pakistanis.

There are, however, evident differences from other South Asian groups. The first is the length of their stay in the West, at least in Britain. This has enabled that particular community to evolve certain strategies for survival (for example, community infrastructures for teaching religion, for social interactions and helping the needy, a body of religious literature and teaching considered appropriate for the new situation). They tended to come to Britain with a distinctive range of perceptions regarding their good relations with the British and their own importance in British India, and in British politics, that made them positive about settling here (not all of them, however, think that Britain has lived up to their expectations). They were unusual in bringing their families with them at an early state of settlement, and they have a far higher incidence of education and professional career patterns than is common among peoples from the subcontinent. In America, of course, most South Asian migrant communities are typically well-educated professionals because of immigration requirements so that the Zoroastrians there do not stand out as much as in Britain, though even by American-Asian standards, Parsis (more than Iranian Zoroastrians) have a very high proportion in the 'executive' class. In Australia, the general pattern of immigration is more like that of America (that is, of people qualified according to the nation's perceived needs, mostly the educated) so that there, also, Parsis stand out as less distinctive than in Britain, yet still even more educated and more successful in business than the other Asian communities in the continent (in proportion to their overall numbers).

Relative to other groups Parsis probably keep less close links with the old country, because for many of them the 'real' old country is Persia, a country with which they cannot currently establish relations. They are, therefore, in some ways a community adrift. The lack of a strong sense of homeland outside Britain, America, Canada and Australia and (originally at least) a positive attitude to relations with the British, probably makes them more ready to identify themselves as British than many other groups – which causes problems of identity for those who consider they face significant prejudice from the general population. In America, Canada and Australia, it is those in the vanguard of migration who appear to be most ready to identify with the country of settlement, but differences have been noted between these countries. The Iranian Zoroastrians are different in whichever country they settle, because of their distinct sense of links to the 'old country'.

Zoroastrians have an unusual, perhaps unique, knowledge that in the 'old country' (be that India or Pakistan) they are a rapidly diminishing community on the verge of extinction, so that the future of their diaspora group is of crucial significance for the very survival of their religion and heritage. In Iran numbers may not be diminishing, but there remains a great sense of vulnerability. The result can be, ironically, more tortuous internal wranglings as different strategies for survival are canvassed. Often these are mutually exclusive yet passionately believed by their upholders to be vital if extinction is to be avoided. They have often been referred to as the Jews of the Eastern world and in their perceptions of the interconnection of religion and race, their memories of centuries of ruthless oppression and persecution, their theological understanding of human nature, destiny and of history, there is much the two communities share.[28]

Different demographic patterns and internal divisions over fundamental issues mean that it is questionable whether one should refer to *a* Zoroastrian community even within a specific locality such as London or New York, much less globally. In the Zoroastrian case divisions are compounded by the national divisions (for example, Zoroastrians have fought on each side in the wars between India and Pakistan and there are, as we have seen, tensions between Zoroastrians from Iran and the subcontinent). As Zoroastrians continue to migrate around the world not only in growing numbers but also, and perhaps more significantly, in a greater proportion compared with the population in the 'old country', so the threat of dispersal seems to make ever closer the apocalyptic scene of the extinction of the world's oldest prophetic religion. Some diaspora Zoroastrians would see that as far too negative an assessment. Many in America and Canada consider that just as the time had come for their religion to move on from its Iranian homeland to India in the eighth to ninth centuries, so now the time has come for it to 'move on' to the 'new world'. They see the community in India as having become too 'Hinduized', as a dying phenomenon, and in Iran as an oppressed and deprived minority whereas in the 'new world' they believe there is a young growing community with the intellectual and managerial leadership to carry Zoroastrianism into the third millennium.[29]

This chapter has suggested that there are many problems not yet properly tackled in the study of diaspora communities, notably the subject of religion. But another conclusion is that it is essential not only to study both ends of the 'chain of migration' (both the old country and the new[30]) but that there is an even greater need for more comparative studies of the diaspora experiences of communities in different countries in the 'new world'. If the Zoroastrians are a reasonable indicator then the diaspora experience is very different from one country to another and in order to see what is distinctive and characteristic of any one country it is essential to compare it with others. Here lies a whole new field of research.

NOTES AND REFERENCES

1. This chapter is based on research conducted between 1984 and 1990 for the Katrak lectures at the University of Oxford, the publication of which is in preparation. The following resources were utilized: (i) the archives of each community listed (minute books of meetings, annual general meetings, correspondence files, etc); (ii) all the books, pamphlets, newsletters and other public documents produced in the various centres; (iii) a survey questionnaire distributed within each community (response rates varied from 30 per cent in East Africa to 95 per cent in Sydney and Hong Kong, yielding a total of 1700 responses). In this short paper it will be possible only to offer a general picture painted with broad brush-strokes. (iv) A programme of 245 interviews was conducted in Britain as a follow-up to the postal survey, mostly by my research assistant Dr Rashna Writer.

   It is both a pleasure and a duty to record my gratitude to the various funding bodies behind this research: the Universities of Hong Kong, Manchester and Oxford; the British Academy; the Spalding Trusts, the World Zarathustrian Trust Funds and individual Parsi sponsors, Tahemton Aresh, Ninoo Treasureywala and the Rivetna family. I learned more from the numerous families I lived with over the years, and from Dr Writer, than from all the records. The computer analysis of the survey data would not have been possible without the help of the whole Mehta family, especially Gillian, and the support of the Manchester University Regional Computer Centre. Archival research would not have been possible without the permission, indeed active encouragement, of the Committees of the Zoroastrian Associations in Chicago, Hong Kong, Houston, Karachi, Lahore, London, Los Angeles, Manchester, Melbourne (Victoria), Mombasa, Nairobi, New York, Sydney, Toronto, Vancouver (British Columbia), Zanzibar. In Bombay I worked mainly in the offices of the Bombay Parsi Panchayet, Zoroastrian Studies, Cama Oriental Institute, Bombay University Library, of various political and social bodies (e.g. Parsi Association, Western Liberal Federation, Masina Hospital) and received help from the editors and records of the magazine *Parsiana*, the newspapers *Kaiser-i-Hind*, and *Bombay Samachar*. But above all I am indebted in India to my friend and patient collaborator, Dastur Dr K.M. Jamasp Asa, High Priest of the Anjuman Atash Behram, Bombay, for his advice, guidance and help in untold ways for 20 years.
2. The best single-volume history of Zoroastrianism is M. Boyce, *Zoroastrians: their religious beliefs and practices* (Routledge, 1979) with an associated volume of translated texts covering the various periods, *Sources for the Study of Zoroastrianism* (Manchester University Press, 1984). There is an extensive commentary on publications relating to the Parsis in J. Hinnells, 'The Parsis: a bibliographical survey', *Journal of Mithraic Studies*, III, 1980, 100–49.
3. See J. Hinnells, *Asians in Britain: the Zoroastrian Experience*, (in press), ch. 2 and meanwhile M.S. Commissariat, 'The first Parsi in England (1724–25); Nowroji Rustom Manek of Surat and his relations with the East India Company', *M.P. Khareghat memorial Volume* (editor not stated) (Bombay, 1953) I, 221–58.

4. See J. Hinnells and J. Pavri, *The Parsi community in China*, (forthcoming) and J. Hinnells, *The Modern Zoroastrian Diaspora*, Katrak lectures, OUP (in preparation). See also K.N. Vaid, *The Overseas Indian Community in Hong Kong* (University of Hong Kong, 1972).
5. The most informed account of the early history in Karachi is J.F. Punthakey, *The Karachi Zoroastrian calendar* (English translation by F.H. Punthakey, Karachi, 1989). The demographic picture as seen in the 1970s is related in E.F. Gustafson, 'A demographic dilemma: the Parsis of Karachi', *Social Biology*, xvi, 115–27.
6. The only items published on Aden are the biography of Hormusjee by A.N. Joshi, *Life and Times of Sir Hormusjee C. Dinshaw* (Bombay, 1939) and an article in the popular Parsi magazine, *Parsiana* (January, 1977), pp. 18ff.
7. Until the publication of the Katraks the only published material on East African Parsis is C. Salvadori, *Through Open Doors: A View of Asian Culture in Kenya* (Nairobi, 1990), pp. 22–8. See also the unpublished manuscript by H. Kased (now of Bombay) *Parsee Lustre on the Emerald Isle of Zanzibar.* I am deeply indebted to Mr Kased for generously giving me a copy of his work.
8. See J. Hinnells, 'The Parsis in Britain', *Journal of the Cama Oriental Institute* (Bombay) [*JCOI*], 46, 1978, 65–84; 'Zoroastrian migration to Britain: its history and impact', World Zoroastrian Souvenir Issue, *JCOI*, 1986, 3–24; and 'Zoroastrians in London' in R. Ballard (ed.), *Desh Pardesh* (in press).
9. J. Hinnells, 'Anglo-Parsi commercial relations in Bombay prior to 1847', *JCOI*, 1978, 46, 5–19; and 'Bombay: the Zoroastrian community' in *Encyclopaedia Iranica*, IV, 4, 339–46.
10. See E. Kulke, *The Parsees in India* (Freiburg, 1974); R.P. Masani, *Dadabhai Naoroji: The Grand Old Man of India* (London, 1939); M. Squires, *Saklatvala: A political biography* (London, 1990).
11. See collected letters in R.P. Patwardhan, *Dadabhai Naoroji Correspondence*, vol. II, parts i and ii (Bombay, 1977). (No other volumes have been published.)
12. *Poverty and Un-British Rule in India*, a collection of his writings on the subject republished by the Government of India (Delhi, 1962).
13. See the personal memoir of his daughter, Sehri Saklatvala, *The Fifth Commandment: Biography of Shapurji Saklatvala*, private publication (Salford, 1991).
14. J. Pavri, 'Brief summary of Zoroastrians abroad' in *Proceedings of the Fourth North American Congress* (Montreal, 1982), pp. 88–95.
15. Source of information Mr R. Wadia, in whose restaurant Parsis tended to meet.
16. For a mere extensive discussion of influences on Dhalla and his theology see J. Hinnells, 'Modern Zoroastrian philosophy' in B. Carr and I. Mahalingam (eds), *Asian Philosophy*. (Routledge, in press).
17. M. Karkal, *Survey of Parsi population of Greater Bombay – 1982* (Bombay, 1984); for Pakistan see Gustafson, note 5 above.
18. Studies of other East African Asians have drawn similar conclusions. See, for example, P. Bhachu, *Twice Migrants: East African Sikh Settlers in Britain* (London, 1985) and V. Robinson, *Transients, Settlers and Refugees: Asians in Britain* (Oxford, 1986), ch. 9.

19. See, for example, K. Mistree, *Zoroastrianism: An Ethnic Perspective* (Bombay, 1983), chs. 12 and 16. Mistree was born in Bombay, but after a professional career studied Zoroastrianism at Oxford and London before returning to Bombay where he set up an influential teaching body, 'Zoroastrian Studies' (ZS). One of ZS's many activities is organizing lectures and seminars. These have proved attractive (but also controversial) to many in London (and America and Canada), especially to the educated youth questing for their roots. His book has, therefore, a particular relevance to the present discussion.
20. Working totally independently of each other, Raymond Brady Williams came to the same conclusion as those outlined above regarding other South Asian groups in these same two cities: *Religious Immigrants from India and Pakistan: New Threads in the American Tapestry* (Cambridge, 1988).
21. See J. Hinnells, 'Zoroastrian migration to the American continent', in M. Treasureywala (ed.), *Proceedings of the 6th North American Congress* (Toronto, 1989), pp. 19–49.
22. See Huzan Karas, 'Intermarriage among Bombay Zoroastrians', PhD thesis, University of Manchester, 1993.
23. An exception is Kim Knott, *Hinduism in Leeds* (Leeds, 1986).
24. See J.J. Modi, *The Religious Ceremonies and Customs of the Parsees* (Bombay, 1937); and Mistree (note 19 above).
25. A good example of such an explanation is Ava Mehta, *The History of our Religion, Zoroastrianism* (Bombay, 1988) pp. 47ff. Although published in Bombay it was written (and delightfully illustrated with cartoons) by a London Parsi schoolteacher for children in the community.
26. See, for example, the exposition of the high priest, Dastur F.M. Kotwal, in Kotwal and Boyd, 'The Zoroastrian *paragna*', *Journal of Mithraic Studies*, II (i), 1977, 35–8.
27. For a study of the various trends and influences, with full bibliography, see Hinnells in note 16 above.
28. Demographic studies of diaspora Jewish communities often raise issues of fascinating comparability to those of Parsi groups. An example I have found instructive is S.M. Cohen, *American Modernity and Jewish Identity* (New York, 1983).
29. Australian Zoroastrians have not yet, to my knowledge, evolved such a perception of their place in Zoroastrian history. Few Zoroastrians outside Hong Kong see the community in that colony as having a long-term future.
30. A recent example of such a dual focus is M. Banks, *Organizing Jainism in India and England* (Oxford, 1992).

4 The Indian Diaspora in the United States

Roger Daniels

Although the overwhelming majority of the nearly one million persons of Indian birth or ancestry presently in the United States have either come in the last quarter century or are the children of those immigrants, the Indian presence there stretches back to nineteenth-century pioneer precursors.[1] Unlike the other major stream of Indian migrants to the New World which brought large numbers of indentured labourers to the circum-Caribbean, all of those who came to the United States (and Canada[2]) came as free persons.[3] No Indians are known to have been recruited by North American employers. By the early twentieth century, when Indian labourers first began to arrive in British Columbia and the American West Coast, the major labour recruitment phase of Asian immigration to North America had ended. That phase had brought hundreds of thousands of Chinese and Japanese workers to the mainland and Hawaii and provoked an anti-Asian movement which eventually all but halted the immigration of Asians to both the United States and Canada for several decades.[4] Although Indians had not participated in the labour migration to North America which produced the nativist anti-Asian legislation, they suffered under it during the decades of little or no migration when Indians, along with other Asians, were ineligible for naturalization and denied other basic civil rights. In the United States the struggle against that legislation, seen by some as a kind of parallel to the freedom struggle being waged in India, was a salient aspect of Asian Indian community life.[5] Indian American lobbyists achieved an important victory in 1946 when Congress dropped all legal discrimination against 'natives of India', but far more important for the future of Americans from India – known as Asian Indians from the late 1970s[6] – was the general liberalization of immigration law in both the United States and Canada.

Although a few Indian seamen were reported in North Atlantic ports from the end of the eighteenth century[7] and there were occasional Indian merchants resident in such cities as New York and Philadelphia in the mid-nineteenth century, it was only in its last decade that Indians made any kind of an impact in the United States. Initially that impact was intellectual and not demographic. It began with a Hindu missionary, Swami Vivekananda. His appearance at the World's Parliament of Religions held in conjunction with the 1893 World's Fair in Chicago was the beginning of what has been a minor aspect of the religious history of the United States: the development of 'Hindu' religious societies whose members are almost exclusively middle class and formerly Protestant Euro-Americans.

To be sure, Vivekananda was not the first such religious leader to come to the United States. Protap Chunder Mozoomdar had come as early as 1884, and, as Carl Jackson has shown, the roots of American sympathy with Indian religious thought can be traced back to the seventeenth century.[8] Such nineteenth-century American thinkers as Ralph Waldo Emerson and Henry David Thoreau had been influenced by Indian thought, while the latter became an important influence on M.K. Gandhi. Other Indian religious influences came via Western exponents and interpreters of Indian religious thought such as Helena Petrovna Blavatsky (1831–91) and the Theosophy movement. Before Vivekananda left the United States in 1895 he had founded the Vedanta Society in New York City. The trip to America apparently showed him the possibilities for making 'conversions', for on his return to India he founded the Ramakrishna Mission. On a second American visit he brought two monks with him who were left in charge of growing Vedanta societies in New York and San Francisco. In the first three decades of the twentieth century, 17 swamis came to the US from the Ramakrishna Mission. Its prosperous 'converts' were numbered in the hundreds by 1930.

Even more successful in numbers of 'converts' was Swami Yogananada who first came to the United States in 1920 to attend the Pilgrim Tercentenary Anniversary International Conference of Religious Liberals. After a successful lecture tour he founded the Togoda Satsanaga Society in Los Angeles. Like the Vedanta Society, the new society emphasized both the philosophy and practice of yoga including exercise and breath control. Yogananada had great talent for both organization and publicity; by 1930 his movement, later renamed the Self Realization Fellowship, claimed 25 000 American followers and had centres in 12 American cities.[9] The most important secular leader to visit North America was the Nobel laureate, Rabindranath Tagore, who came several times beginning in 1916. His influence may be seen in the establishment of Tagore societies with a membership of both ethnic Indians and others in New York and California.[10]

None of these movers and shakers had anything directly to do with the Indian labour migrants who entered the North American West in the early years of this century. There is no clear beginning to their migration. The United States Census of 1900 reported 2050 persons born in India, but almost certainly this is inaccurate. American immigration data show the entry of 5771 natives of India as immigrants between 1900 and 1910, all but 109 of them males. Towards the end of that period standards were tightened as 1618 were barred from entry, almost half of them, 777, on grounds of extreme poverty, or, as the law put it, 'likely to become a public charge'.[11] Canadian data show 5200 entries of Indians between 1904 and 1910.[12] From 1911 to 1920 only 102 Indians are recorded as entering Canada and some 1500 the United States.[13] It is clear that many of these newcomers were counted more than once – most Asian Indians in this era entered the continent through Vancouver – and that some were not counted at all.

To put these numbers into perspective, total immigration from all countries to the United States in the first decade of this century was nearly nine million while almost a million and a half entered Canada. Thus fears of being swamped by what some writers called 'a tide of turbans' were chimerical.[14] But such fears did exist and were a subsidiary part of the anti-Asian reaction which decades of Chinese and Japanese immigration had produced in western North America. An editorialist for the Vancouver *Sun* spoke for the white western majority in both countries when he argued that it was important to restrict even the few Indian immigrants who were coming because: 'There are 300 000 000 natives of India behind them.... if a million were accepted, the remaining 299 000 000 would have the same claim.'[15]

The vast majority of the Indian immigrants who came in the first decade of the 1900s were from the Punjab and most of those were Sikhs.[16] The immigrants were initially concentrated on both sides of the 49th parallel in the Pacific Northwest and worked chiefly on railroads and in lumber mills. The railroad workers were widely dispersed in 'section gangs' but the lumber mill workers, concentrated in cities, drew the fire of white working men. In 1907 there were riots against Asian immigrants throughout the Pacific Northwest.[17] The worst to be directed solely against Indians occurred in Bellingham, Washington, a small industrial city on the eastern shore of Puget Sound 18 miles south of the Canadian border. There seem to have been about 250 Indians there with 153 of them employed at five lumber mills. Labour relations in the lumber industry were notoriously brutal[18] and labour leaders argued that lower paid 'Hindus' were being used to replace white workers. Sporadic violence against Indians began on the Labour Day holiday, Monday, 2 September 1907, and peaked on Wednesday night. After an evening mass meeting at which workers were urged to 'move them on, get them out of town, and scare them so badly that they will not crowd white labor out of the mills', a mob of some 400–500 whites invaded the ethnic enclave where the Indian workers lived and beat many of them, some badly. Happily, none was killed. Most seem to have left for nearby British Columbia or California. Anti-Indian disturbances followed in at least six other American west coast towns.[19] Although most of the press deplored the violence, many editorialists approved its ends. As one Seattle paper put it:

> It is not a question of race, but of wages; not a question of men, but of modes of life.... When men who require meat to eat and real beds to sleep in are ousted from their employment to make room for vegetarians who can find the bliss of sleep in some filthy corner, it is rather difficult to say at what limit indignation ceases to be righteous.[20]

Investigators for the United States government agreed. Its Immigration Commission reported in 1911 that:

> The East Indians on the Pacific Coast are almost universally regarded as the least desirable race of immigrants thus far admitted to the United States ... far below the Japanese, Chinese or other oriental races ... [T]hey are generally the least efficient race employed but they accept employment at low wages, and if admitted in large numbers might seriously [lower wages].[21]

Bruce La Brack, one of the chief students of Indian immigrants, has described the early arrivals as 'single, Punjabi Sikhs, predominantly Jat peasant-cultivators. Most were labourers, albeit skilled agriculturalists, who had little if any formal education, [and] spoke little English'.[22]

As noted above, new interpretations of existing immigration law in the United States, coupled, as will be seen, with new regulations in Canada, severely limited the entrance of Indians in the years before World War I. In 1917, following the so-called 'barred zone' Act, immigration of Indians to the United States was halted for more than three decades, although students, merchants and religious leaders could still enter.[23]

Canada effectively stopped Indian immigration with an Order in Council in January 1908 which gave immigration officials the authority to bar any would-be immigrants to Canada who did not come on a 'continuous voyage' from their native country. Since there was no direct steamship service from India to Canada – most immigrants had come via Hong Kong – Indians could be and were barred without the embarrassment of any act or regulation singling them out (Canada and India were, after all, in the same Empire).[24] Six years later in the spring and early summer of 1914, 376 would-be Indian immigrants to Canada, all but 30 of them Sikhs, staged a two-month sit-in in Vancouver harbour on board a Japanese vessel, the *Komagata Maru*, which had been chartered in a vain attempt to breach the immigration barriers.[25]

While no lives were lost in the incident itself its aftermath was bloody. Indian militants in both Canada and the United States were spied upon and harassed by an Anglo-Indian policeman, William C. Hopkinson, borrowed by the Canadian government for that purpose. His chief agent in the Vancouver immigrant community was a Sikh named Bela Singh. Two of Bela Singh's agents were murdered. During a post-funeral service in the *gurdwara* (Sikh temple, literally 'gateway of the guru'), Bela Singh killed two men and wounded six others. At his trial Hopkinson, who had volunteered as a character witness for the defence, was shot to death by Mewa Singh, the priest of the *gurdwara*. Mewa Singh was unrepentant,[26] and was hanged on 11 January 1915.

The virtual cessation of Indian immigration to North America left tiny, largely male, Indian communities in four locations: in British Columbia around Vancouver and in the Okanagan Valley,[27] and in California's Sacramento and Imperial Valleys. It is clear that most, if not all, of those who became permanent settlers had originally come intending to make some money and return to India, which is what many did. La Brack argues that it is wrong to think of these men as individual pioneers because most seem to have emigrated 'as the outcome of a family corporate decision' and travelled with 'others who stood in some kinship relation to them'.[28] According to one contemporary observer, a common goal was to save US$2000. Wages were low. Even the most frugal workers could expect to save no more than US$25 a month. At that rate it would take more than seven years to amass US$2000, so most of the would-be sojourners were bound to be frustrated. But savings did occur. The post office at Marysville, California, reported US$34 000 in remittances to India in one nine-month period during 1908–9. The one contemporary survey, based on interviews with 70 millhands in 1909, reported that 31 of them had sent a average of nearly US$140 to India the year before.[29]

Those in agriculture usually worked in gangs, often under the leadership of one who knew the ropes and had a better command of English. Western farmers had been dealing with similarly-led groups of Chinese and Japanese labourers, and immigrant European workers in both agriculture and industry also utilized similar arrangements. The gang bosses were often the first entrepreneurs in the immigrant communities and among the first to lease or buy land.[30]

Few and poor as they were, the Sikhs on the West Coast established lasting cultural institutions. In 1908 Vancouver's Sikhs established the first *gurdwara* on the continent. Two years later Sikhs in the San Francisco Bay area founded the Pacific Coast Khalsa Diwan Society (Free Divine Communion Society) which built the first *gurdwara* in the United States in Stockton in 1915. This temple became, figuratively if not literally, the 'mother house' for American Sikhs. How these establishments were financed is not clear; the Vancouver temple supposedly cost US$6000.[31] Some of the funds may well have come from Sikhs outside North America; for example, Gurdit Singh, a Hong Kong entrepreneur, chartered the *Komagata Maru*. By the end of the decade the Khalsa Diwan Society had also established a residence in Berkeley for Indian students at the University of California.

By the time of World War I – a time of great prosperity for California farmers – some Sikhs had begun to lease and buy farmland in the Imperial and Sacramento valleys. (This apparently did not happen in Canada's Okanagan Valley until the 1930s). By 1919, when the state of California conducted a survey of 'oriental' landholding, an estimated 2600 'Hindus' owned or leased nearly 90 000 acres of farmland, most of it in either cotton or rice. On a per capita basis the Sikh farmers controlled more than six times as much land as did Japanese farmers there, but the Japanese, of course, were much more numerous.[32]

The previously mentioned barred zone immigration act of 1917 meant that any who travelled to India probably could not return. The population did decline; those who went back probably were not proprietors. The census, notoriously inaccurate for small minority populations, reported only 1873 persons born in India in 1930 and 1476 in 1940. These were obviously no longer sojourners, but permanent residents.

All but a handful of the permanent residents of Indian birth were male; although many speak of the bachelor society of Asian migrants of this era, many of the Sikhs, like other Asian migrants, were not really single. Rather, they were permanent residents without women. Many were married, with wives and children in India. A sizeable proportion, perhaps a majority, of the Indians who became agricultural proprietors, entered into American marriages, many of which were bigamous under American law. La Brack and Karen Leonard have reconstructed nearly 400 marriages of California Indian farmers before the resumption of Indian immigration in 1946. Only nine of these – 2.4 per cent – were with Asian Indian women. About 80 per cent were with Hispanic women, most of them from Mexican and Mexican-American farm worker families.

These Punjabi-Mexican marriages were conflict-ridden, with an extraordinarily high percentage ending in divorce. In Imperial County, for example, at least one in five of such marriages was formally dissolved; husbands and wives instituted proceedings in roughly equal numbers.

Frequently the Hispanic spouses were women who had worked for their future husbands and in most marriages the wife was many years younger than the husband. Often she moved into an otherwise male household consisting of her husband and his Punjabi partners. There was a tendency for sisters to marry partners so that many households contained related women and, sometimes, their parents and other distaff relatives.

Surviving wives speak of their husbands' 'single-minded concentration' on putting all their resources into farming. Some immigrant husbands were also sending remittances back to India to support wives and children there. The strains and stresses of these intercultural marriages can be seen in the divorce petitions.

The men's complaints stress neglect of duty by the wives, particularly a refusal to clean and cook for the husbands' friends, verbal disobedience, too much visiting by mothers and sisters, too much shopping, using make up, and dancing with other men, usually Mexicans or Mexican-Americans. The women's complaints stress their husbands' drinking and demands for unreasonable domestic services. Both sets of complaints report violence, verbal abuse including racial and ethnic slurs and sexual infidelity. It must have been a great shock to the first Punjabis who underwent divorces to discover that under California's community property laws their wives could receive a share of both real and personal property, child custody and support and even alimony.[33]

On the other hand most of these marriages seem to have endured. Neither religion nor language seem to have been major issues of disagreement. Few of the men had significant formal education and none were priests. Their culture was not transmitted to their children. Some of the first daughters born to these couples were married at an early age to Punjabi men a generation or two older, but these marriages usually ended in divorce. Leonard has investigated 220 marriages of the children of Asian Indian-Mexican marriages, only 11 of whom married another Punjabi-Mexican.[34] Had American immigration law not been changed this process of 'circuitous assimilation'[35] would have continued and the Punjabi ethnic group in California would have disappeared.

One other Asian Indian group in North America during the early decades of the twentieth century needs to be considered: the several handfuls of student and other radicals who organized the Ghadar Party during World War I, and dreamed, foolishly, of overthrowing the British Raj in India from a base in California by using funds supplied by the Imperial German government to ship arms and revolutionaries back to India. (Some of those involved in the *Komagata Maru* affair and its aftermath were, or became, Ghadarites). Compromised by informers and agents provocateurs, conspirators were arrested in both the United States and Canada and many of the revolutionaries who reached India were killed one way or another without ever striking a blow. Those tried and convicted in San Francisco received only short sentences – for violating the neutrality acts – as the United States was not at war when the plotters were arrested; those convicted in Canada got longer terms. There were homicides connected with the San Francisco trial. One accused Ghadarite shot a co-defendant to death right in the courtroom and was himself killed by a US marshal.[36] Mere rationality cannot explain the Ghadar syndrome. As the translated words of a contemporary protest song put it:

> The time for prayer is gone;
> It is time to take up the sword.
> Empty talk does not serve any purpose.
> It is time to engage in a fierce battle.
> Only the names of those who long for martyrdom will shine.[37]

In the quarter-century after World War I the Asian Indian community in North America shrank steadily. Many of those Indians who remained were or became prosperous. A few score merchants and other middle-class individuals came, largely as non-immigrant residents. Despite the ban on the naturalization of Asians, a number of Asian Indians had been awarded American citizenship in various federal courts. The naturalization statute of 1870 had limited naturalization to 'white persons' and 'persons of African descent'. The Chinese Exclusion Act of 1882 had specifically forbidden the naturalization of Chinese, but there was some doubt about other ethnic groups. A test case from Honolulu begun in 1914 was finally decided by the Supreme Court in 1922. A long-time acculturated educated Christian Japan-born resident, Tadeo Ozawa, conceded to be 'well qualified by character and education for citizenship', was nevertheless denied it on the grounds of race by a unanimous court. When asked to define the statute's vague term 'white', the court answered that white meant 'Caucasian'. This seemed to open the door for another litigant, the Sikh, Bhagat Singh Thind, who although quite dark, was ethnologically a Caucasian. His case came before the court in 1923. Without a single dissent the justices reversed themselves. Agreeing that Thind was a Caucasian, the court now held that 'white' in the 1870 statute was to be construed 'in the understanding of the common man'.[38] The decision had repercussions throughout the Indian American community.

At the time of the Thind decision there were perhaps 70 naturalized Asian Indians scattered across the United States. In the next three years the Department of Justice managed to get some 50 of these naturalizations revoked. But in 1926 an Asian Indian immigrant, Sakharam Ganesh Pandit, who had received his naturalization certificate in 1914 and had subsequently been admitted to the California bar, took the Government to court and argued, successfully, that since it was an established rule of law that a decision not appealed within three years should stand, his naturalization could not be revoked.

This result assisted the efforts of another middle-class immigrant, Prafulla G. Mukerji. Mukerji, after participating in nationalist activities in Calcutta as an undergraduate, emigrated to the United States in 1905 with just US$50. He worked for a year as a hydraulic engineer in New Jersey, was graduated from the University of Pittsburgh in 1911 and then worked for the US Steel Corporation until his retirement in 1956. He had been naturalized sometime after 1915. In 1924 he received a letter from the Attorney General of the United States instructing him 'to show cause' why his citizenship should not be revoked. Seeking political rather than legal recourse Mukerji appealed to his senator, David Reed (Republican, Pennsylvania), who introduced a congressional joint resolution permitting Asian Indians holding citizenship to keep it. It died in committee, but because of the Pandit case Mukerji's citizenship was not revoked. Not content with that, he organized what may have been the first concerted attempt by Asian Indians to influence Congress: he helped establish the tiny Hindu Citizenship Committee which sought to obtain the right of citizenship for all Asian Indians.[39]

The Thind decision also put California farmers at risk, for it confirmed that they were 'aliens ineligible to citizenship' and thus subject to state alien land laws which barred such aliens from owning or leasing agricultural land. There were ways to get around these laws, the most common of which for Asian Indians was marriage to someone not covered by the law. But it did underline the fact that there was considerable prejudice against Asian people in California.[40] Not surprisingly the successful Indian farmers in California also began to lobby for citizenship through such organizations as the Hindustan Association of America.[41]

That this lobbying effort would succeed seemed far-fetched. Anti-Asian prejudice in the United States was well entrenched and the Asian Indian community was tiny. But by the time of World War II and its immediate aftermath, the intense Cold War era of the 1950s, citizenship for immigrant Asians was an idea whose time had come. The first breakthrough came for Chinese in 1943. After a well-crafted lobbying campaign by prominent

Caucasians and with the strong endorsement of President Roosevelt, Congress repealed the Chinese Exclusion Act, made Chinese, but no other Asians, eligible for naturalization and gave them an immigration quota of 105 per year.[42] This was presented as a foreign policy measure: China was an ally in the war against Japan. Roosevelt told Congress that he regarded the legislation in removing anachronistic laws from the statute books as 'important in the cause of winning the war and of establishing a secure peace' and in strengthening alliance relations.[43]

Despite FDR's rhetoric, the exclusion of Chinese was not an anachronism but an integral part of a restrictionist policy that had barred more than half of the world's people from either immigration or citizenship. We can see now, although it was not apparent at the time, that the repeal of Chinese exclusion in 1943 was the pivot on which all American immigration policy began to turn, the hinge on which the 'golden door' of American immigration began to swing open again. Three years later the door was opened a bit wider for 'natives of India' and Filipinos. In 1952, as Cold War hysteria moved toward its apex, all racial and ethnic bars to naturalization and immigration were removed. The restrictionist and essentially isolationist immigration policy had been at odds with the internationalist foreign policy of Roosevelt and his successors. During World War II it had been important to accommodate one's allies; during the Cold War, when the United States aspired to global leadership, it was important not to alienate potential recruits to the 'free world'.

Even so, the fact that India, not yet independent and with nationalist leaders who had not been supporters of the war effort, was rewarded with citizenship at the same time as the Philippines and six years before it was granted to the rest of Asia, requires explanation. It can be partially explained by the long American anti-colonialist tradition, however dubious its pedigree might be. Millions of Americans regarded Gandhi as a hero and the future of India was one of the issues that divided the wartime leaders, Roosevelt and Churchill.[44] But without acknowledging the lobbying efforts of a number of Indians resident in the United States it is impossible to explain why the door was opened for Indians *in 1946*. Despite the prior campaigns of several Indian organizations, the indispensable lobbyist was a Sikh merchant, long resident in New York, who called himself 'J.J.' Singh.

Nearly five years after his great triumph, Robert Shaplen glowingly profiled Singh in the *New Yorker* magazine:

> Sirdar Jagjit Singh, the president of the India League of America, a privately sponsored nonprofit organization that seeks to interpret India

> to, and to further Indian causes in this country, is, at fifty-three, a handsome, six-foot Sikh, who by means of persistent salesmanship, urbane manners, and undeviating enthusiasm, has established himself as the principal link between numberless Americans and the vast mysterious Eastern subcontinent where he was born.[45]

Singh operated in the classic American lobbyist style: he cultivated members of Congress, influential opinion-moulders and the press. He attracted support from a galaxy of Americans all across the political spectrum, from Roger Baldwin of the American Civil Liberties Union on the left to press baron William Randolph Hearst on the right. The bill he was promoting – granting both the right of naturalization and an immigration quota[46] – was first introduced in 1944 by a bipartisan odd couple, Democrat Emanuel Celler and Republican Clare Booth Luce.[47] It became law in 1946 and provided that 'persons of races indigenous to India' were eligible for naturalization and established an annual quota of one hundred persons for India.[48] The act caused a mild sensation in India – Gandhi, for example, wrote an enthusiastic note to Singh about it – but its immediate demographic consequences were slight. US immigration data show fewer than 7000 'East Indian' immigrants between 1948 and 1965. On the other hand this immigration probably at least doubled the Indian American population and greatly reduced its gender imbalance. Almost 6000 of them were classified as 'non-quota', mostly close relatives of residents.

Singh, a patriot who saw his campaign as part of the struggle to free India, never applied for American citizenship. A California Sikh, Dalip Singh Saund, was among the majority who did. Born in Amritsar in 1899, he was the son of a well-to-do but illiterate government contractor. The Jallianwalla Bagh massacre of 1919, which occurred while he was taking a degree in mathematics at Punjab University, shocked him into leaving India after graduation. He came to the United States via England and entered the University of California at Berkeley where he took three degrees – an MA and PhD in mathematics, and, for practical reasons, an MS in agricultural science. While at Berkeley he lived, rent free, in the Khalsa Diwan's student 'clubhouse' and was active in the Hindustan Association of America, becoming its national president. Turning down two offers of professorships from Indian universities, he decided to remain in the United States. As he tells it he soon found there was no place in American academia for him[49] and 'that the only way that Indians in California could make a living was to join' with compatriots in agriculture.

More than a year after his graduation – a year not detailed in his autobiography – he went to work as a foreman on a cotton ranch in the Imperial

Valley operated by fellow Indians. He soon became a rancher himself and prospered. He also took on more intellectual tasks such as publishing a response to Katherine Mayo's notorious *Mother India* (1927) under commission (subsidy?) from the Sikh temple in Stockton. Like many but not all upwardly mobile Sikhs in North America in those years he was determined to acculturate – he had begun shaving and stopped wearing a turban shortly after he emigrated – and in 1928 he married the daughter of an upper middle class Czech-American family. He and his wife engaged in a wide range of civic activities, many of which involved freedom for India and civil rights for Indian Americans. With the egotism typical of politicians, his three-page account of 'the fight for citizenship' ignores Jagjit Singh and most of the other leaders and organizations that helped to win it.

Shortly after becoming a citizen Saund was elected a local judge and then in 1956 won a seat in Congress as a Democrat. Taking office in January 1957 he became the first of many Asian American congresspersons. Saund's election attracted much attention and was exploited as a Cold War ploy by the United States information Agency which sent him on a tour of Asia – among other activities he addressed the Indian parliament – as evidence of ethnic democracy in the United States. He was twice re-elected, but lost in 1962 after a stroke confined him to a hospital bed.[50]

There are no reliable estimates of Indian American population in the 1960s as the census did not enumerate Asian Indians separately until 1980. For 1970 one scholarly estimate put the Asian Indian population at 75 000 and in 1980 the census counted 387 223.[51] As late as 1987 skilled demographers projected an Asian Indian population of just over a million for the year 2000, but the 1990 census total of 814 000 shows that this is too low an estimate.[52]

To understand how this massive increase came about it is necessary to note the major revamping of American immigration law in 1965. The National Origins System, set up in 1924, was scrapped and a two-tiered system substituted for it. One tier established annual ceilings of 170 000 for the eastern hemisphere and 120 000 for the western, and set an annual maximum of 20 000 for each nation, although, obviously, each nation could not have 20 000. The second tier was family oriented. Spouses, unmarried minor children and parents of US citizens – native born or naturalized – were exempt from any numerical limitation and entered in addition to those in the first tier; and even unnaturalized permanent residents could get spouses and children in under a high preference but subject to numerical limitation. Between 1976 and 1985, for example, the number of relatives admitted without numerical limitation ranged from

113 000 to 210 000 annually. In addition, a whole host of special refugee programmes have been superimposed upon the 'regular' immigration so that total legal immigration to the United States has grown from 3.2 million in the period 1961–70 to 7.3 million between 1981–90.[53]

It is not just numbers that have changed; the sources of immigration have changed as well. Europe, which once contributed the lion's share of immigrants to the United States, contributes an ever smaller part, while immigration from Asia and Latin America has become dominant. Between the 1950s and the 1980s the European share fell from 52 per cent to 11 per cent, while immigration from Latin America grew from 25 per cent to 42 per cent, and that from Asia from 6 per cent to 42 per cent. In actual numbers immigration from Asia grew from an annual average of 13 300 in the 1950s to 268 500 in the 1980s.[54] The following tables show the growth of the Asian American population and of its major ethnic groups.

Table 4.1 Asian American population, 1950–90

1950	599 091
1960	877 934
1970	1 429 562
1980	3 466 421
1990 (preliminary)	7 272 662

Table 4.2 Asian Americans, 1990, by major ethnic and national groups

Chinese	1 643 621
Filipino	1 403 624
Japanese	850 901
Asian Indian	814 538
Korean	799 993
Vietnamese	610 904
Laotian*	239 000
Cambodian	210 724

*Includes Hmong

Not just the number but the incidence of Asian Americans has been increasing. The 7.3 million Asian Americans of 1990 represent almost 3 per cent of the American population, up from 0.7 of 1 per cent as recently as 1970, a more than threefold increase in 20 years. Asian Indians, if we accept the 75 000 estimate for 1970, have more than doubled their incidence among Asian Americans in the same period, from 5 per cent to 11 per cent.

Asian Indians were particularly well positioned to take advantage of the changes in immigration law, which, as we have seen, favours family reunification. This, in turn, leads to chain migration, migration in which related individuals follow one another. Family reunification after 1946 led to a more balanced sex ratio. Increasingly available and rapid air transportation made it easier for Indians who could afford it to come and go; improved telephone and later facsimile facilities made intercontinental communication relatively simple. As soon as an immigrant with a permanent visa arrived in the United States, he or she (but among Asian Indians usually he) could arrange for a spouse and minor children to get visas under a high preference. And as soon as an immigrant was naturalized, spouses and minor children could come in without numerical limitation and the new citizen's adult children and siblings could get preference numbers. These new adult permanent residents could then begin the whole process again.

Elliott Barkan's statistical study of recent American naturalization patterns shows that (1) Asians become citizens faster than any other large group, and (2) that among them Asian Indians showed the greatest propensity for rapid naturalization, which Barkan defines as five to eight years after establishing residence. During 1969–78 65 per cent of Asian immigrants naturalized rapidly, as opposed to 45 per cent of all others. Among Asian Indians the rate was 80 per cent; Filipinos, the next highest Asian group, were recorded at 60 per cent.[55]

Since at this writing (March 1993) no detailed analysis of the 1990 census data is available, the following statistical portrait of the Asian Indian community is based on the 1980 data. There is good reason to believe that most of the observed patterns are continuing.[56] Unlike its predecessors the contemporary Asian Indian community is urban, not rural, and it chiefly results from family migration, not that of men unaccompanied by women.[57] Marriages in the new immigrant communities are not only endogamous, but tend to be within ethnic subcultures. No longer concentrated in the Far West, Asian Indians are widely distributed: just over a third live in the Northeast, nearly a quarter each in the South and Midwest, and fewer than a fifth in the Far West.

Perhaps the most remarkable aspect of Asian Indian family life in the contemporary United States is its stability: 92 per cent of all Asian Indian households contained both husband and wife and 92.7 per cent of all children under 18 years of age were in two-parent families. This is higher than for any other ethnic group identified by the census.[58] Asian Indian families were not large; Asian Indian women aged 25–34 had produced 1336 children per 1000, more than similarly aged women of most other

Asian American groups, but below the US average of 1476. As is usually the case among immigrant groups, the more recently arrived households were larger and contained more unrelated persons than did those longer established.

The adult members of these households are highly educated, and their children, whether born in India or in the United States, are continuing that tradition. More than nine out of ten Asian Indian males, aged 25–29, are high school graduates as are nearly nine out of ten similarly aged females: they outstrip all other groups identified in the census except Japanese Americans. For the same age groups, almost two-thirds of the males and just over half of the females have completed four or more years of college.

As one would expect for such an educated population the data for income and occupation show that Asian Indians were relatively well-off and engaged in essentially middle-class occupations. Income figures for 1979 – which include income from wages and salaries only – show Asian Indian males earning US$20 643 annually, as opposed to US$18 759 for whites. For Asian Indian females the gap was wider: they averaged US$9685 as opposed to US$7908 for whites.[59] Almost two-thirds of the Asian males reporting occupation were in jobs the census bureau classified as 'executive, administrative or managerial' (14.9 per cent), 'professional' (41.7 per cent), or 'technical' (8 per cent). The corresponding figure for whites was just over a quarter, while for blacks it was nearly a seventh. Just over two-fifths of the employed female Asian Indians were in the same three categories, as opposed to just over a quarter of white females and nearly a fifth of black females.

An examination of professional occupations shows some of the specialization within the Indian community. More than two-fifths of male professionals were engineers and over a quarter physicians. Among female professionals better than three-fifths were in health-diagnosing occupations.

In addition many Asian Indians were entrepreneurs.[60] In most large American cities, tandoori or other Indian restaurants are the most visible sign of this, although in some ethnic neighbourhoods sari shops are prominent. There is an ethnic logic in these businesses, but not in two other niches that Asian Indians have come to occupy disproportionately, motels and (petrol) stations. But both are ideal for immigrants practising chain migration in that each is a labour-intensive enterprise with relatively small capital requirements. Family, extended family and clan members perform much of the semi-skilled labour, and some new arrivals, after learning the trade and raising capital, establish similar businesses of their own. So much of the Asian Indian motel business is in the hands of Gujaratis who share the surname Patel that a community joke speaks of 'hotel, motel, Patel'.[61]

In New York City perhaps 40 per cent of all petrol stations are now operated by South Asians. One successful entrepreneur recently told his story to the *New York Times*. Parmjit Singh came to New York in 1981 at age 21. After working as a bus boy and cook in a restaurant for three years he met another Punjabi who owned a petrol station. 'He taught me how to run a station', Singh recalled. 'It's very hard work but it's easy to run – to order gas, to clean, to keep the books'. By 1990 he owned 13 stations. The Punjabi entrepreneur, who reports that his stations net about US$4000 a month, employs Sikhs almost exclusively. These men earn about US$300 a week and save half of it. 'They see me and they want to be like me', he told the *Times* reporter.[62]

Indian community leaders are convinced that unlike most immigrant groups in the United States Asian Indians will maintain most of their cultural traditions and continue to practise endogamous marriage. They expect Indian immigrants and their children to continue to do well in American society, but not acculturate in other ways. The degree to which their unique religious heritage will insulate them and their children and their children's children remains to be seen. Most students of immigration and ethnicity in America would expect successive generations of Asian Indians to follow patterns of acculturation similar to those of other Asian immigrants.[63]

One of the curious aspects of the development of such a large Indian American community so quickly is its relative anonymity. If one compares the attention given in the press to Asian Indians with that given to Koreans – whose number and rate of growth are quite similar – it is easy to see why some authors speak of the 'invisibility' of Asian Indians. There are a number of reasons for this – two of which are their relative non-concentration in California and the Far West and the fact that many of the practitioners of Asian American studies deliberately ignore them.[64] Asian Indians have not been prominent in ethnic civil rights coalitions since 1946. Recently some public hostility has been directed against them, hostility that seems to come largely from lower-class whites rather than from blacks, as has been the case with Koreans. The worst violence has been in Jersey City where there have been sporadic but organized assaults on Indians by groups that call themselves 'dot busters', a reference to the *bindi*, the small cosmetic dot that married Hindu women traditionally wear on their foreheads. At least one Indian, a 30-year-old man who was a manager in the credit card division of a New York bank, died after a beating. The middle-class Asian Indians are not 'stealing' the jobs of whites, but their very industry, propensity to save rather than consume, and relative affluence clearly create jealousy among their less prosperous white neighbours. Real-estate purchases and the resultant gentrification in once all-white neighbourhoods were clearly a factor.[65] While a far cry from the Bellingham riot of 1907 – today's press is universally condemnatory and most public officials, especially New Jersey Governor Thomas Kean, expressed strong support for the Asian Indian community – the Jersey City incidents were merely one of a number of cases of anti-Asian violence scattered all across the United States. Despite success, despite being hailed as 'model minorities', Asian Americans still faced at least residual discrimination.[66]

Notwithstanding these and other difficulties – difficulties undoubtedly worsened by seemingly endemic recession and sluggish economic growth in the United States – it is possible to be relatively sanguine about the immediate future of the Asian Indian community. Even if immigration were to be cut off again – not a likely event – the community has, by the 1990s, a large enough population base to assure its survival and expansion for the foreseeable future.

NOTES AND REFERENCES

1. The major study is J.M. Jensen, *Passage from India: Asian Indian Immigrants in North America* (New Haven, 1988), which treats the early twentieth century. For Canada see N. Buchignani et al., *Continuous Journey: A Social History of South Asians in Canada* (Toronto, 1985); H. Johnston, *The East Indians in Canada* (Ottawa, 1984); and S. Chandrasekhar (ed.), *From India to Canada* (La Jolla, Calif., 1986). J. Singh et al. (eds) *South Asians in North America: An Annotated and Selected Bibliography* (Berkeley, 1988) is the best bibliography. G.R. Hess, 'The Forgotten Asian Americans: the East Indian Community in the United States', *Pacific Historical Review*, 43 (1974) 577–96 is a pioneering summary. See also R. Daniels, *History of Indian Immigration to the United States; An Interpretive Essay* (New York, 1989) and the chapter 'Asian Indians', in H.H.L. Kitano and R. Daniels, *Asian Americans: Emerging Minorities* (Englewood Cliffs, NJ, 1988). I have been enlightened by a paper by Surendra Bhana, 'Asian Exclusion in Comparative Perspective' and profited from his comments on a draft of this chapter.
2. Because Indian emigration to North America was essentially one stream in the early twentieth century there will be some account of Canada here. After both nations began to regulate Indian immigration the streams became discrete.
3. The classic account is H. Tinker, *A New System of Slavery: The Export of Indian Labour Overseas, 1830–1920* (London, 1974). For a general account of Sikh migration see S. Chan, 'Sikhs Overseas in the Context of International Migrations', in M. Juergensmeyer and N.G. Barrier, (eds) *Sikh Studies: Comparative Perspective on a Changing Tradition* (Berkeley, 1979)

and N.G. Barrier and V.A. Dusenberry (eds), *The Sikh Diaspora: Migration and the Experience Beyond Punjab* (Columbia, Mo., 1989).

4. R. Daniels, *Asian America: Chinese and Japanese in the United States since 1850* (Seattle, 1988).
5. P. Varma, 'The Asian Indian Community's Struggle for Legal Equality in the United States, 1900–1946' (PhD thesis, University of Cincinnati, 1989).
6. Confusion, in the census and elsewhere, between natives of India and their descendants and aboriginal Americans and their descendants, a confusion begun by Columbus, caused community activists to petition the Census Bureau to create the 'Asian Indian' category, which it did in time for the 1980 census. Prior to that time the term 'East Indian' was used. In this essay the term 'Indian' will be used for both persons born in India and their American-born descendants.
7. Jensen, *Passage from India*, pp. 12–13.
8. C. Jackson, *The Oriental Religions and American Thought: Nineteenth Century Explorations* (Westport, Conn., 1981). Jackson will soon publish a study of the Ramakrishna Mission.
9. W. Thomas, *Hinduism Invades America* (Boston, 1930).
10. See S.K. Mukherjee, 'Passage to America: The Reception of Rabindranath Tagore in the United States, 1912–1941' (PhD thesis, University of Pennsylvania, 1963) and Q.N. Uddin, 'Horizon of Expectations: The Reception of Rabindranath Tagore in the United States and Britain (1913–1941) (PhD thesis, State University of New York, Binghamton, 1985).
11. Other major causes were: trachoma, 347; surgeon's certificate of defects which might affect the ability to earn a living, 179; loathsome or contagious disease, 120; and contract labourers, 79. U.S. Immigration Commission, *Reports of the Immigration Commission*, vol. 23, Part III, 'The East Indians on the Pacific Coast', pp. 323–49, tables 1, 2 and 5.
12. J.G. Chadney, 'India's in Vancouver: Immigration, Occupation and Ethnic Adaptation'. Table 1, p. 60. in Chandrasekhar, (ed.), *From India to Canada.* The 1908 order stopping Indians from entering Canada is described below.
13. B. La Brack, *The Sikhs of Northern California, 1904–1976* (New York, 1988), table 2, p. 78.
14. J.R. Lewis, '"The Tide of Turbans" – North American Images of Punjabi Immigrants in the Early 20th Century', *Punjab Past and Present*, 21 (1987) 87–102.
15 As cited in W.P. Ward, *White Canada Forever: Popular Attitudes and Public Policy Toward Orientals in British Columbia* (Montreal, 1974), p. 90.
16. For general conditions see I. Ali, *The Punjab under Imperialism, 1885–1940* (Princeton, NJ, 1988).
17. For details see H. Sugimoto, *Japanese Immigration, the Vancouver Riots, and Canadian Diplomacy* (New York, 1978) and R.E. Wynne, *Reaction to the Chinese in the Pacific Northwest and British Columbia, 1850 to 1910* (New York, 1978).
18. N. Clark, *Mill Town: A Social History of Everett Washington* (Seattle, 1970).
19. G.N. Hallberg, 'Bellingham, Washington's Anti Hindu Riot', *Journal of the West*, 12 (1973) 163–75.
20. Seattle *Morning Times*, 7 September 1907. The use of dietary differences to illustrate cultural differences was well established. Samuel Gompers, for

example, had published an anti-Chinese pamphlet entitled *Meat vs Rice*. For a 'liberal' American view in a national magazine see 'Hindu, the Newest Immigration Problems', *Survey*, 1 October 1910.

21. U.S. Immigration Commission, *Reports*.... 23,iii, 'East Indians...' p. 349.
22. La Brack, *Sikhs of Northern California*, p. 81
23. The 'barred zone' act created an area, described in degrees of latitude and longitude, whose 'natives' could not immigrate into the United States. This included India and almost all of Asia. See 39 *Stat*. 874, Sec. 3.
24. For details see Buchignani et al., *Continuous Voyage*; and Ward, *White Canada Forever*, ch. 5.
25. H. Johnston, *The Voyage of the Komagata Maru: The Sikh Challenge to Canada's Colour Bar* (Delhi, 1979). The ship sailed from Hong Kong to Vancouver and thus fell foul of the 'continuous voyage' rule.
26. K. Singh, *A History of the Sikhs Volume 2, 1839–1964*, (Princeton, NJ, 1966), pp. 178–180. According to Chadney: 'The martyrdom of Mewa Singh is celebrated in the *gurdwara* every March and is a major event on their ceremonial calendar'. *The Sikhs of Vancouver*, p. 28.
27. A. Joy, *Ethnicity in Canada: Social Accommodation and Cultural Persistence Among the Sikhs and the Portuguese* (New York, 1989), is an inadequate treatment of the Okanagan communities.
28. La Brack, *Sikhs of Northern California*, p. 96.
29. U.S. Immigration Commission, *Reports*..., 23, iii, 'East Indians...' p. 349.
30. A.P. Miller, 'An Ethnographic Report on the Sikh (East) Indians of Sacramento Valley' MSS., South Asia Library, University of California, Berkeley. For immigrant entrepreneurship see V.R. Greene, *American Immigrant Leaders, 1800–1910* (Baltimore, 1987).
31. Buchignani et al., *Continuous Journey*, p. 24.
32. California State Board of Control, *California and the Oriental* (Sacramento, 1922), pp. 26, 47 and *passim*.
33. B. La Brack and K. Leonard, 'Conflict and Compatibility in Punjabi-Mexican Families in Rural California, 1915–1965', *Journal of Marriage and the Family*, 46 (1984) 527–37. K. Leonard, *Making Ethnic Choices: California's Punjabi-Mexican Americans* (Philadelphia, 1991) is a major contribution.
34. K. Leonard, 'Marriage and Family Life Among Early Asian Indian Immigrants', *Population Review*, 2 (1982) 67–75.
35. Y. Dadabhay, 'Circuitous Assimilation Among Rural Hindustanis in California', *Social Forces*, 33 (1984) 138–41.
36. Jensen, *Passage from India*; M. Juergensmeyer, 'The Ghadar Syndrome: Nationalism in an Immigrant Community', *Punjabi Journal of Politics*, 1 (1977) 1–22; and J. Singh, 'The Gadar Party: Political Expression in an Immigrant Community', *South Asia Bulletin*, 2 (1982) 29–38.
37. V. Prakash and S. Vatuk, 'Protest Songs of East Indians on the West Coast, U.S.A.', *Folklore*, 7 (1966) 370–82.
38. *Ozawa v. U.S.* 260 *U.S.* 189 (1922); *U.S. v. Thind* 261 *U.S.* 204 (1923). The original 1790 naturalization statute had limited the process to 'free white persons'.
39. M.P. Fisher, *The Indians of New York City: A Study of Immigrants from India* (Columbia, Mo., 1980), pp. 13–16.
40. For further detail on the land laws see, D.S. Saund, *Congressman from India* (New York, 1960), pp. 45, 64–5.

41. Saund, *Congressman*, pp. 68–96. Varma, 'The Asian Indian Community', is the fullest account of the citizenship struggle.
42. The act was 57 *Stat.* 600. The quota applied to 'Chinese persons', not citizens of China so that a Chinese citizen of another country – say Britain – who emigrated to the United States, would need one of the 105 quota spaces for that year. However, American immigration law provided that certain close Chinese relatives of United States citizens could enter without reference to the quota.
43. S.I. Rosenman, (comp.), *The Public Papers and Addresses of Franklin D. Roosevelt, 1928–1945*, 13 vols. (New York, 1938–50), 1943 volume, pp. 429–30.
44. W.R. Louis, *Imperialism at Bay: The United States and the Decolonization of the British Empire, 1941–1945* (Oxford, 1978).
45. Shaplen, 'One-Man Lobby', *The New Yorker*, 24 March 1951, pp. 35–45. See also R. Narayanan, 'Indian Immigration and the India League of America', *Indian Journal of American Studies*, 2, no.1 (1972) 1–30; M.S. Venkataramani & B.K. Shrivastava, *Quit India: The American Response to the 1942 Struggle* (New Delhi, 1979); and Varma, 'The Asian Indian Community'.
46. There was a tactical division among ethnic group leaders in the United States. Some, like G.J. Watamull, a prosperous California businessman, would settle for 'half a loaf'–that is naturalization for those already here – but no further immigration.
47. E. Celler, *You Never Leave Brooklyn: The Autobiography of Emanuel Celler* (New York, 1953).
48. Act of 2 July 1946, 60 Stat. 416. Much racism remained embedded in American law. Some pedant inserted three pettifogging paragraphs to delineate which quota should be charged for a person of 'one-half Chinese blood and one-half the blood of a race indigenous to India'.
49. For nativism in the American mathematical profession see N. Reingold, 'Refugee Mathematicians in the United States, 1933–1942: Reception and Reaction', in J.C. Jackman and C.M. Boardman, (eds), *The Muses Flee Hitler: Cultural Transfer and Adaptation, 1930–45* (Washington, DC, 1983), pp. 205–32.
50. Saund, *Congressman from India*, and Saund, *My Mother India*, (Stockton, Calif., 1930).
51. H. Elkanialy and R.W. Nichols (eds), *Immigrants from the Indian Subcontinent in the U.S.A.: Problems and Prospects* (Chicago, 1976).
52. L. Bouvier and A. Agresta, 'The Future Asian Population of the United States', in J.T. Fawcett and B.V. Caniño (eds), *Pacific Bridges: The New Immigration from Asia and the Pacific Islands* (New York, 1987).
53. For a good brief summary of how the law has operated see L.W. Bouvier and R.W. Gardner, *Immigration to the US: The Unfinished Story* (Washington, DC, 1986).
54. See the chart in W.P. O'Hara and J.C. Felt, *Asian Americans: America's Fastest Growing Minority Group* (Washington, DC, 1991), figure 3, p. 6.
55. E.R. Barkan, 'Whom Shall We Integrate? A Comparative Analysis of the Immigration and Naturalization Trends of Asians... (1951–1978)', *Journal of American Ethnic History*, 3 (1983) 29–57. Five years is the minimum residence period for citizenship.
56. Much of what follows is adapted from P. Xenos et al., *Asian Indians in the United States: A 1980 Census Profile* (Honolulu, 1989).

57. However some of the original Asian Indian rural enclaves have been re-enforced since 1965. See especially M. Gibson, *Accommodation Without Assimilation: Sikh Immigrants in an American High School* (Ithaca, 1988) and 'Punjabi Orchard Farmers: an Immigrant Enclave in Rural California', *International Migration Review*, 22 (1988) 28–50.
58. The next highest reported figure was for Koreans, 89.4 per cent. 'Whites' had 82.9 per cent, Hispanics, 70.9 per cent and blacks, 45.4 per cent. Gardner et al., *Asian Americans*, table 8, p. 21. In many US census compilations, Hispanics are listed twice: first tabulated as either white or black (most are counted white), and then aggregated separately. This tends to make white families appear less stable and larger, white income lower, etc., than would be the case if Hispanics were not included.
59. Xenos et al., *Asian Indians*, table 17, p. 34. These figures overstate average income because 'persons with a reported income of zero dollars are excluded'.
60. For a celebration of this see A. and U.M. Helweg, *An Immigrant Success Story: East Indians in America* (Philadelphia, 1990).
61. Little has been written about the low-wage employees of some of the ethnic enterprises, but for an impressionistic sketch see S. Thaker, 'The Quality of Life of Asian Indian Women in the Motel Industry', *South Asia Bulletin*, 2 (1982) 68–73.
62. D. Lorch, 'Ethnic Niches Creating Jobs That Fuel Immigrant Growth', *New York Times*, 12 January 1992. Similarly Afghans operated more than 200 fast-food chicken franchises.
63. See H.H.L. Kitano et al., 'Asian American Interracial Marriage', *Journal of Marriage and the Family*, 46 (1984) 179–90; L.H. Shinagawa and G.Y. Pang, 'Marriage Patterns of Asian Americans in California, 1980', in S. Chan (ed.), *Income Status Differences Between White and Minority Americans* (Lewiston, NY, 1990), pp. 225–82; P.R. Spickard, *Mixed Blood: Intermarriage and Ethnic Identity in Twentieth Century America* (Madison, 1989). R.B. Williams, *Religions of Immigrants from India and Pakistan: New Threads in the American Tapestry* (Cambridge, 1988), is a learned examination of immigrant religious practices.
64. The lack of concentration in California is relative, not absolute. In Artesia, a suburb of Los Angeles, there is a four-block stretch that some now call Little India, which is probably the third largest Asian Indian ethnic enclave after Jackson Heights in Queens (New York City) and Devon Street in Chicago.
65. A.A. Narvaez, 'Jersey City Indian Community Protests Rash of Racial Acts', *New York Times*, 8 October 1987; M. Marriott, 'In Jersey City Indians Face Violence', ibid., 12 October 1988; M. and S. Jain, 'Indian Success in Depressed North Jersey Rouses Jealousy', *India Tribune* (Chicago), 21 November 1987; S.S. Hedge, 'Prepare for the Challenge', *India Tribune*, 21 November 1987; and private information.
66. See, for example, two publications of the United States Commission on Civil Rights: *Recent Activities Against Citizens and Residents of Asian Descent* (Washington, DC, 1986) and *Civil Rights Issues Facing Asian Americans in the 1990s* (Washington, DC, 1992).

5 Renewing an Industrial Past: British Pakistani Entrepreneurship in Manchester

Pnina Werbner

INTRODUCTION

The growing literature on ethnic and immigrant entrepreneurship in Europe and the United States raises important questions regarding the differential mobility of various ethnic groups,[1] and the implications of business success for an understanding of certain forms of racism.[2] It also raises questions regarding the cultural basis of immigrant business success, starting from a Weberian emphasis on the unique cultural features of a capitalist 'ethos' and moving to a critique of this approach which underlines the collective resources that immigrants mobilize, both ethnic and class,[3] and their unique placement as immigrants and sojourners.[4] Much of this literature stresses the solidaristic dimensions of immigrant entrepreneurship.

General questions regarding the relationship between class, disadvantages in the labour market and immigrant entrepreneurship have also been widely considered.[5] Some authors raise doubts regarding the potential of immigrant business, particularly in Britain, to create a viable economic base in a depressed economy.[6] Marxist approaches regard most immigrant business as a disguised form of cheap labour exploitation.[7] A recent innovative development has been Wilson and Martins' discussion of immigrant 'enclave economies', echoing anthropological studies of economic niches.[8] My own work[9] also stresses the special features of the immigrant enclave economy and its potential for capital expansion. Some of these theoretical issues within the broader field of ethnic and racial studies will be addressed here.

MANCHESTER: THE EMERGENCE OF TRADING AND MANUFACTURING ENCLAVES

In its heyday during the nineteenth century Manchester had a 1000 warehouses in the central city, alongside 99 spinning mills, dyeing and printing works.[10] Inner-Manchester districts such as Ancoats and Cheetham Hill had a sweatshop sector built on a long tradition of outwork. By the 1950s, however, Manchester was a 'deindustrializing' city with a declining population and a contracting labour force. The textile industry had all but collapsed, with employment in 1981 only 10 per cent of that in 1951. The centrifugal forces sending both industries and people away from the centre hit the inner city particularly hard. Industrial employment declined by 43 per cent. Moreover, Mellor argues, 'the small firm sector which had characterised the core since the nineteenth century was collapsing'. By 1976, 72 per cent of Manchester's employed population were in the service sector.[11]

Within this context of general industrial and commercial contraction the social economy of Pakistani immigrants represents a remarkable counter-trend.[12] Like the Chinese and other Asian immigrants, they settled in a declining industrial city which was fast losing its prominence even as a regional metropolis. They have shifted during this 30 year period of decline into manufacturing, wholesaling, and distribution, primarily in the food and garment sectors. To facilitate this move, they utilized empty warehouses and workshop space, creating new businesses in the abandoned landmarks of Manchester's former industrial glory.

The flourishing of immigrant entrepreneurship in an apparently declining economy raises general theoretical questions. If, for example, we are to argue that immigrant entrepreneurship must be understood as responsive to favourable 'opportunity structures',[13] then the Manchester case illustrates that such 'opportunities', if they exist, must nevertheless be perceived, recognized or 'discovered'. In other words, the construction of meaning crucially determines economic strategy. Manchester, a city apparently in inexorable decline, has become for Asian immigrants a base for potential economic prosperity.

Manchester was throughout the nineteenth century the international commercial centre of the textile industry. Part of this industry focused on the import and export of greycloth (i.e. unprinted cotton). This sector of the trade was dominated by foreign merchants primarily from the Levant, many of Jewish origin. On a lesser scale, the city was also a centre of clothing manufacturing and wholesaling. In the pre-war period the main sectors of manufacturing were rainwear, outdoor wear and haberdashery (especially the manufacturing of cloth caps), and these industries were dominated by Jewish immigrants from Eastern Europe, who had arrived in Manchester in the early and especially latter half of the nineteenth century. Jewish firms formed an integrated network of small businesses specializing in 'shmeering' (a process of rubberizing rainwear garments discovered in the nineteenth century). Most of these firms have disappeared following technological innovations in the manufacture of rainwear during the 1950s and the diminishing demand for cloth caps.

Jewish immigrants were also involved in large numbers in tailoring, the sewing of made-to-measure suits. Alongside manufacturing was market trading, and from the interviews I conducted this followed a similar pattern to more recent British Pakistani market trading: Jewish market traders lived in Manchester and traded peripatetically throughout the North-West. The most successful traders became wholesalers and many of the large High Street chain stores had their origins in small-scale wholesaling (most prominently, Marks and Spencer which started in Manchester, and developed from market trading, retailing and wholesaling). High Street retailing appears to have been followed by a diversification in the manufacturing sector, especially into ready-to-wear clothing commissioned by the chain stores, but Manchester retained, on the whole, its specialized national monopoly of rainwear manufacturing and haberdashery.

In line with the general decline since the war, and particularly following the influx of imported outdoor wear from the Far East and changing fashions, these peculiarly Manchester Jewish specializations in rainwear and cap cloth manufacturing have also all but disappeared.

Hence, against the trend towards industrial decline, incoming immigrants from India and Pakistan have emerged as a new, and increasingly significant, entrepreneurial class.[14] Although they are involved in a vast array of small businesses, including the food and catering industries, hotels, taxis, retailing, etc., they have concentrated, like Jewish immigrants before them, in the clothing and garment industries/trades. The specialization in rainwear and haberdashery has been replaced by a concentration on the manufacturing of ready-made clothing (mainly trousers, skirts and co-ordinates) for markets, and most recently, of cheap acrylic knitwear, a major development in the city. During the post-war era, market trading, wholesaling and manufacturing at the cheap end of the trade were taken over initially by Jewish refugees fleeing from Europe who utilized contacts with established Jewish manufacturers and fabric suppliers to found their businesses. They established themselves more quickly than the incoming South Asians (who arrived in numbers somewhat later, during the 1950s). Recent Jewish immigrants became suppliers to Pakistani market traders, and customers of local manufacturers. However, over the years competition from incoming Asians increased and virtually all of these Jewish wholesalers have now sold out to them. The clothing sector of the trade, and particularly the cheap end of it, is today dominated by South Asian immigrants. There are over 500 firms in the city, both wholesaling and manufacturing, which sub-contract to a wide range of further firms within the North-West, possibly numbering several thousand, and service large numbers of retailers, predominantly market traders, whose markets cover a large region within a radius of 100 miles or more from the city. The vast majority of these firms are in the hands of British Pakistanis.

VISIBILIZATION

The dominance of Asian trading, initially somewhat hidden (since most traders were market traders operating from home) has now become tangibly evident in the city landscape. An arc of wholesaler firms now stretches immediately adjacent to the High Street retailing district for about a mile and a half (Map 5.1). Its city-centre sector specializes in women's clothing and fashion at a medium price level, while its northern sector specialises in casual wear, jeans, children's wear, outdoor clothing and cheap electrical and fancy goods. This northern sector, known as Strangeways, was the traditional manufacturing-cum-wholesaling district dominated in the past by Jewish tailors and other firms. With three giant wholesalers now dominating this area, most cash-and-carry shops are small, but their number is constantly increasing.

Interspersed with and surrounding the wholesale district is a further arc of small-scale manufacturers, who mainly concentrate in specific locations, especially in areas of converted warehouses and old mills. My door-to-door survey of the wholesale and cash-and-carry district showed a vast preponderance of South Asian owners, with a few Jewish firms (mainly in ladies' fashionwear) and rare firms owned by members of other groups (such as Greek Cypriots). Although most firms have neutral or English names, the owners are predominantly South Asian, the majority Muslim (Muslims predominate in the city as a whole). Some of the very largest wholesalers are, however, Hindu, and this has important implications for an understanding of the changing relationship between trading and community.

The textile trade, like the clothing trade, has undergone a shift in the post-war era, from cotton export to cotton import, primarily from South Asia. This sector of the trade, as mentioned, tended to be dominated by 'foreigners', primarily originating from the Middle East and Iran. Thus many Sephardic Jews from Syria, Egypt, Iraq, Iran, the Lebanon and most recently India, as well as Christians and Muslims from these countries, concentrated within this trading enclave. From 1948 onwards the textile trade began to incorporate growing numbers of South Asian textile merchants, and they have become increasingly successful as the direction of trading has shifted. The Jewish

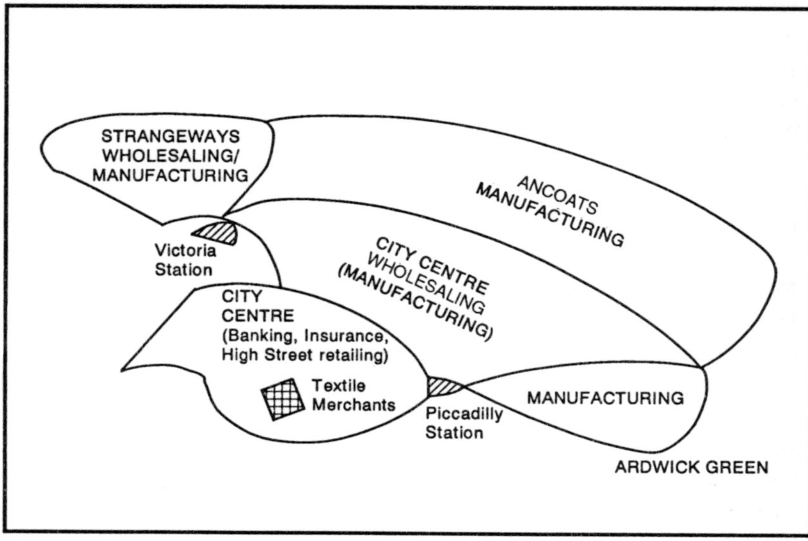

Map 5.1 Asian wholesaling and manufacturing in Manchester

merchants were hit by the closure of Middle East markets and by the growing decline in the demand for British textiles. This transition benefited the Asian newcomers, well placed to tap the growing Indian and Pakistani cotton manufacturing industry. Some of the more successful Asian textile merchants have expanded vertically into cloth conversion and other related businesses, with the largest forming a finance company.

There is a fundamental difference between the clothing/garment trade and the textile trade in the way both capital and knowledge are utilized. The textile trade is highly capitalized but no merchant can stock the full range of cottons the market demands. Moreover, temporary shortfalls in capital require mutual guarantees, usually provided within circles of mutually trusting merchants. Hence, relations of trust and credit are paramount in the textile trade and essential for its smooth operation. It is a gentlemen's trade: merchants work from offices equipped with a telephone, telex and fax machines. Social relationships between merchants sustain at all times a façade of friendship and generosity, disguising competition. By contrast, the clothing trade is highly competitive and uncertain; it involves constant hustling and short-term deals. Textile merchants deal in what are essentially commodities. Thus, knowledge is a prerequisite for competent trading and has a measure of certainty in it. Clothing traders deal in luxury items with uncertain values. Textile merchants are interdependent horizontally; clothing traders are interdependent vertically. The divide has social connotations in the mutual perceptions traders and merchants have of one another: the merchants inevitably regard themselves as benevolent members of an exclusive circle of genuine businessmen, unlike the fly-by-night, untrustworthy traders; the traders regard the merchants as proud and status-conscious. This mutual stereotyping is reinforced by the fact that traders and merchants tend to belong to different subethnic groups. A social divide exists today between Punjabi clothing traders (both Hindus and Muslims) and predominantly Gujerati or Marwari merchants (primarily Hindu but including some Muslim Mehemanis from Karachi). While the divide is not absolute, it marks basic orientations and lifestyles.

The majority of the South Asian immigrant-controlled firms in the city are of relatively recent origin, and with few exceptions they started with very little capital. All the firms, even the larger, more successful amongst them, may be classified as small firms. They vary, however, in size from very small firms with a relatively low turnover of well under half a million pounds per annum, to larger firms with sales of up to 10 or 15 million pounds per annum. While very few of the firms (possibly only one or two) are mentioned in any Company annuals, the differences in size are highly significant if seen from the perspective of Asian entrepreneurs themselves, both within the trade and, indeed, in the various Asian communities. One of the most important features of the two enclaves (clothing and textiles) is that both are organized hierarchically. This is undoubtedly true of any business or industry, but the small scale of immigrant businesses by national standards means that this hierarchy is almost entirely hidden from the eyes of business analysts. The fact that there is a hierarchy, and that it is perceived as such, emerged in the course of the fieldwork itself. Before going on to discuss this hierarchy in detail, let me pursue first the question of the origins of Asian entrepreneurship in Manchester, with reference to the general sociological problem raised in most studies of immigrant entrepreneurship: to what extent has Asian entrepreneurial success in Manchester stemmed from immigrant solidarities of specific 'ethnic' competences, and to what extent was it a consequence of class 'resources'?

THE FORMATION OF AN ENTREPRENEURIAL BASE

The research revealed a complex pattern of entrepreneurial growth. Most striking was the length of time it took to build up a base sufficient for the type of ethnic entrepreneurial 'take-off' currently evident.

Initial migration to Manchester was of several caste and ethnic subgroups originating from different social locations and backgrounds. The resulting historically located social divisions have come to be, in the space of a generation, cut across by other segmentary and horizontal divisions within the local Pakistani and Indian communities, and this has facilitated the current broad-based entrepreneurial movement evident among Asians in Manchester.

Asian migration to Manchester may be regarded as composed of parallel yet distinct migratory movements or flows.[15] For a while, during the late 1950s and early 1960s, the various movements appeared as one. Yet in reality, the arriving migrants followed distinct migratory patterns.

A well-documented feature of Asian migration to Britain has been the pre-war arrival of migrants from the Punjab who became door-to-door pedlars. It is noteworthy that these migrants settled in the larger cities: in London, Birmingham, Manchester, Leeds, Glasgow or Newcastle. The question of why they became pedlars is not usually raised. In an interview with the son of one of these early pedlars in Manchester, he explained that his father and the cohort of migrants he belonged to had few employment options available to them. The British economy was in deep recession and there were no jobs open to these migrants. In Manchester such pedlars are known to have arrived in the 1930s and most continued to struggle for many years, operating on a tiny scale until the war and the early 1950s. Ironically, however, it is this no-option depressed economy which the early pedlars encountered that set the basis for the economic fortunes of the Asian community in Manchester.

Very generally the early 'trader' migrants included both Punjabi Hindus and Muslims. My information is more detailed about the Muslims, but my interviews indicate that the pattern was true for both religious groups. Muslim Punjabi traders came from three kinship-cum-caste groups. The earliest, and much the largest group, was of members of the Arain caste from Jullunder. They followed each other to Manchester through a process of chain migration, and once they had established themselves in the clothing trade (initially as pedlars, later as market traders, wholesalers and manufacturers), they incorporated newcomers very rapidly into the expanding niche. In its initial stages this group relied heavily on 'immigrant' solidaristic support. They had no experience in business since the majority were of village origin. It is also noteworthy that they were assisted during the 1950s by Jewish wholesalers (themselves pre-war or post-war refugees), as well as by members of their own community. The earliest wholesalers were Muslims, but they were joined after a brief period by some Hindu wholesalers. (It is remarkable that whereas most of the prominent Muslim traders were East Punjabi by origin, several of the prominent Hindu traders were West Punjabis. Both groups thus became refugees at the time of Partition.) Two other kinship-cum-caste groups arriving about the same time did have some trading experience. One extended family was composed of itinerant petty traders in the Punjab, and the kin group extended into Singapore and Malaysia. After Partition members of this kindred, of the Rawal *biradari* (brotherhood), settled in Wazirabad, Gujranwala, while at the same time some began migrating to Manchester. They too entered successfully into market trading and shop-keeping. One prominent branch of the family owns a medium-sized hotel, another is a large wholesaler. Some are in the grocery trade.

A third *biradari*, also with international trading links in the Far East, came somewhat later, in the 1950s. Many members of this kindred were tailors by occupation. They too originated from Gujranwala. Given their expertise several became successful manufacturers, while a few are whole-salers as well.

Members of all three groups and especially the two Gujranwala *biradaris*, were literate and relatively educated to high school or college level, and even beyond. This, despite the fact that most came of village or small-town origin. None, however, had any start-up capital. They were all Punjabis and they incorporated newcomers from their groups into trading and manufacturing after very brief periods in factory wage-employment. Hence a combination of factors precipitated the move into entrepreneur-ship: a highly disadvantaged labour market; a very high degree of immigrant solidarity based on kinship and caste rather than simply ethnic links, and the tiny size of the immigrant population (a few hundred). These combined in some cases with some prior trading experience or class resources, the latter a more prominent factor in the further expansion of the trade.

The second major migrant flow to Manchester was that of students. Like other major British cities, post-war Manchester attracted a large number of students to its educational institutions. Education at that time was free, visas were not required. The most successful of these students completed their studies and returned to Pakistan or India to occupy choice positions within the emerging elite there. Many students, however, remained in the city, either to practise (as accountants, doctors, etc.) or to go into business. Like the migrant traders, most Muslim students were born in India, and most were Punjabis although there were some Urdu speakers from Delhi, Lucknow, and other parts of the United Provinces, and a few Gujarati speakers from Karachi. The families of most of these students had moved into the cities of Pakistan after Partition. Virtually all belonged to higher castes, and while forming a sub-group, many had connections to the 'trader' East Punjabi migrants. In any case, the community in those days was very small, and concentrated residentially around the university,[16] all of which made for links between the two groups. Hence, the second major group to become entrepreneurs were these ex-students, often because they had not completed their degrees or faced disadvantage in the open labour market (especially when it came to promotions and salaries).

As the trade expanded, and as some immigrants became wealthy, migrants from other areas, primarily Muslim Punjabis from West Pakistan who had come as factory workers, joined the trade. These migrants were not connected to the trading families and only a small minority attempted initially to move into market trading, shopkeeping or manufacturing. Like the earlier pedlars, however, this enterprising minority was to lay the foundation for this group's entry into business in the 1980s. The movement between generations has meant a further move by the 'workers' and their children into the trading community, and by some of the children of both established traders and workers into professional occupations. Young men's networks cut across earlier divisions by area of origin and this facilitates their entry into market trading. Many young men also work in Asian-owned factories, essentially going through an apprenticeship period before opening their own factories, usually in partnership with their fathers and other family members. In other words, the foundation laid by the early pedlars and traders created a base for entrepreneurship for groups beyond their immediate family members and *biradaris*. The 'entrepreneurial chains' which were generated reached out to other Asian subgroups through ties of friendship, neighbourliness, employment and common associational activities.

In other respects, too, the early divisions have been cut across by emergent divisions. Hence, while differentials in wealth within the community have increased, they do not simply reflect the migratory flows identified. Some traders have prospered while others continue to struggle. Some of the workers who entered into manufacturing and market trading in the late 1970s have done remarkably well within a relatively short space of time. Their move into trading and business has gathered momentum with the maturation of their families and the greater availability of family labour.[17] Their networks beyond the city allow them to mobilize clothing machinists – always a scarce resource – within a wider region.

This movement into the trade by 'workers' can be regarded as utilizing collective 'ethnic' resources built up in the course of several decades of Asian entrepreneurship. The nature of this entry is, however, not as straightforward as it might appear: the barriers to entry must be breached, and in the discussion below this aspect of 'learning' or knowledge-acquisition will be analysed. Moreover, the influx of newcomers, and especially of British-educated younger Asians, into the trade has been associated with a breakdown of trust in the field of credit extension. This also has implications for the way the trade operates.

THE 1980s: HIERARCHY AND EXPANSION

During the earlier period, until 1981, the Asian sector of the clothing and garment trade/industries was dominated by a single large Punjabi Muslim firm with a turnover unparalleled even now by most of the large wholesalers. The firm expanded into importing and into multiple retailing and, given its very large turnover, had acknowledged hegemony over all other Asian clothing firms in the city. In 1981, during the severe recession which hit the trade, this firm overextended itself and collapsed. No other Muslim firm rose in its place to the same position of dominance. Instead, several Hindu Punjabi firms replaced it as the largest importers and wholesalers in the city. This altered the relation between community and trading, especially within the Pakistani Muslim community. An exception to this Hindu hegemony is a single Muslim clothing firm (Joe Bloggs) which expanded enormously during the 1980s and is still growing. The owner of this firm is, however, an acknowledged 'outsider' in the Pakistani community. He moved to Manchester quite recently from an outlying small town in Lancashire where he had a retail shop, the basis of his current fortune. He was born in Lucknow, originates from Karachi, is non-observant, and appears to foster close social ties with the dominant Hindu wholesalers, being one of the few Muslims to attend their weddings and major social gatherings on a regular basis. He does have a circle of Muslim friends within the community, primarily non-observant like himself, but he does not involve himself in communal affairs.

Seen financially and organizationally, the hierarchy in the trade may be divided into several levels: at the top are the three major wholesale establishments, two Punjabi Hindu (headed by related members of the same extended family) and the above-mentioned Muslim. These firms have turnovers of over 10 million pounds a year and are currently expanding fast. Below these three are some 10 major wholesalers, several of them Hindus but the majority Punjabi Muslims, with turnovers of three to five million pounds per annum. Below these are a vast number of smaller wholesalers and cash-and-carry establishments, the majority once again

Muslims, with turnovers of one million pounds and less. The individual circumstances of these firms is usually precarious, and there is a relatively high rate of attrition at this level. Manufacturers without retailing or wholesaling branches rarely have a turnover of more than a million pounds per annum and many have much less than this.

This changing structure of inequality in the trade has meant that there is no single wealthy Muslim wholesaler willing to act also as communal benefactor and philanthropist, as the early wholesaler and the cohort surrounding him had done.[18] Many Muslim traders and manufacturers depend for credit and custom on Hindus and the correspondence between position in the clothing trade and in the community has, to some extent, been severed. The fragmentation of the Manchester Pakistani community and the lack of clear authority structures within it can thus be dated to this alteration in the hierarchical structure of the trade and to the relative independence it gives smaller Pakistani retailers and manufacturers. By the same token, however, successful Punjabi Hindu traders have begun to foster their links with the wealthy Hindu textile merchants to form a wealthy class of Hindu businessmen, and this has encouraged broader organizational unity within the Indian community in the city, transcending the many internal divisions within this community.

The spurt in the fortunes of some business establishments can tentatively be explained by the central role of second-generation sons or younger brothers in the expansion (or collapse) of businesses. The Punjabi Muslim firm which dominated the trade until the 1980s was, in fact, a second-generation firm, developed by the sons of the original owner. Its demise, although precipitated by the serious recession of 1980/81, is attributed, more specifically, to the ill-considered business decisions and purchases of the third generation of sons, as they entered the business. The largest Hindu firm in the city was developed from a retail shop in Liverpool market by the three sons of the original owner, all of them university graduates with MAs in scientific and engineering subjects. The other large Hindu firm is also run by a somewhat older pair of brothers who joined their father's business. The largest Muslim wholesale establishment in the city today was developed to its present size, and is still run in all matters of purchasing and trading, by the owner's very young son. Most of these sons are in their early or late twenties. The story is a common one: almost all the expanding firms are run by second-generation sons of the family. On the whole, but not universally, Hindus appear to be more successful in encouraging their sons to pursue their academic education for longer, and although few appear to study subjects relevant to the clothing trade, I can only conclude that the fact that they are educated affects their performance in the business. The select group of young educated enterprising Muslims who are taking over their families' businesses also have independent social relations with one another, irrespective of the links between their fathers.

This role of the younger generation remains to some extent hidden. It became evident only after I conducted a large number of interviews. I was aware, of course, that the expansion of businesses is related to an expansion in the pool of trusted labour – this is true of all the firms I studied. What was not so evident initially was the qualitative leap that this younger generation seemed to generate, seen in terms of their sense of fashion, sophistication, managerial skills, communicative skills and sheer energy.

Usually sons defer to fathers as having overall authority in the business, and my interviews were thus often conducted with fathers rather than sons. In cases in which fathers had died or retired, sons were available, and only then did it become clear how central the role of this educated new class of businessmen was. The pattern that has emerged is that fathers continue to control long-term financial policies, but most decisions, and the day-to-day running of the firms, is left to the sons.

The hierarchy of the trade is also, very importantly, a hierarchy of credit, not only in terms of the vertical interdependencies of credit extension between retailers, wholesalers and manufacturers which I have discussed already,[19] but also in terms of the amounts of credit extended regularly. Large wholesalers extend large amounts of credit, some regular customers being allowed as much as £20 000 in credit on a regular basis. The vast sums of money tied up in credit must be underpinned by a strong capital base and a substantial operating budget or cash flow. Large wholesalers face the most serious problem in the trade today most acutely: the erosion of trust and creditworthiness, especially in trading relations with market traders. A similar problem is faced by manufacturers who extend credit to wholesalers. Smaller wholesalers experience endemic cash flow problems and are often being run at a hidden loss. The first to suffer when the businesses of these small wholesalers fail are the manufacturers.

Controlling credit and checking creditworthiness is thus a major problem and challenge in the trade today, and often spells the difference between success and failure. Asked why credit defaulting has become so prevalent, most old-timers regarded this development as stemming from an attrition in the values both of society at large and within the Asian (mainly Pakistani) community. Most market traders today are young Pakistanis who have grown up in Britain. The evident wealth of some members of their community spurs them into taking unwarranted risks in their desire to make a quick fortune. The value placed on personal honour and reputation, so important for the fabric of relations sustained in the early immigrant community, has been lost.

Pursuing credit defaulters is an almost hopeless task, given the weakness of the law and the cost of litigation. Most wholesalers and manufacturers tend instead to allow in their budgets for a certain level of credit defaulting (£10–20 000 a year was the sum regularly mentioned), a reasonable, even low sum to write off for a business with a £3 million turnover. Traders, however, naturally resent this type of loss bitterly, and credit defaulting was regularly mentioned in all the interviews I conducted as a serious problem.

THE GROWTH OF THE ENCLAVE AND THE DEVELOPMENT OF BUSINESS PORTFOLIOS

I have written quite extensively about the dynamics of ethnic economic enclave development.[20] Very briefly, I have argued that the enclave grows through a process of horizontal multiplication and vertical extension and that the growth is fuelled by external economic factors on the one hand (recession and disadvantage in the labour market) and internal factors on the other (the maturation of families and the growth in the availability of a trusted labour force). Hence, although increased entry into the trade is associated with intensified competition between ethnic firms, it is also associated with overall growth and the emergence of very large, highly capitalized firms at lower levels of supply (mainly, in this case, importing and wholesaling).

The current research confirmed this picture, but it also pointed to a further dimension of this growth: the formation by firms of complex business portfolios. One major focus of the research was thus to study the expansion strategies of individual entrepreneurs. To do so, the need is to examine the growth of the larger Asian firms which have had time to develop such complex business portfolios. The study of economic portfolios revealed how the two critical parameters of entrepreneurship – capital and knowledge – are interlinked.

It is usually assumed that small firms grow gradually – by expanding the volume of their turnover, the size of their premises, the quality of their products, the sophistication of their technology, and so on. We think of a small firm, in other words, as restricted to single premises and a single line of production. Large centre firms, by contrast, are characterized by vertical and horizontal integration: they contain, within a single firm, many different types of operation, lines of production, premises, etc. When they grow, they do so through acquisition of pre-existing firms and their incorporation into the parent firm. The different branch-firms making up a parent firm supply each other, and funds may be shifted from one company to another to maximize profits and minimize costs.

But are small firms really so radically different from large centre firms? Do they really expand simply through more and better of the same? The evidence from this study suggests otherwise. It was clear that immigrant firms do grow gradually at first. Initial growth takes the form of linear rather than complex expansion. Hence, a recurrent story of most firm owners concerned their initial start as market traders. Market traders often began with one suitcase, then two, travelling by public transport. The next move was to acquire a vehicle, and to shift from a temporary to a permanent stall, then to double the length of the stall. Henceforth the attempt was to find additional stalls in other markets, occupying as many days of the week as possible (markets are usually only held three days a week, and it is particularly difficult to get a permanent stall in a good Saturday market). Successful market traders tend to end up finally with several stalls in separate markets, they become mini-chain retailers, employing several workers at different markets. Most market traders told the same tale.

Clothing manufacturers, like market traders, tended to start on a very small scale, utilizing hired sewing machines and the labour of wives or daughters. The next move was usually into specialized rented premises, often quite small, around 500 square feet. Further expansion was followed by the purchase of machines, bigger premises, the employment of packers and pressers and an increase in the number of machinists used regularly, many of them outdoor workers. Manufacturers who expanded further often brought some of the machinists into the factory. At a later stage they preferred to buy their own premises rather than continuing to rent. In Manchester many of the most established manufacturers now sub-contract to smaller firms in Lancashire and Yorkshire, expanding their production volume once more. This line of expansion means that manufacturers are usually the last to develop highly complex portfolios. Knitwear manufacturers tell a similar tale. A manufacturer starts with a couple of second-hand knitwear machines. As his volume grows he buys newer and more sophisticated machinery, and changes premises. If he does very well he buys computerized knitwear machines.

It is more difficult to enter into textile import–export these days without large reserves of capital, but the story of the textile merchants' growth in the late 1950s and during the 1960s was also initially a story of linear expansion. Starting on a small scale, they gradually increased the volume of goods imported or exported. Based initially in a small office with a telephone and telex (and nowadays fax), they acquired larger offices and a bigger staff of accountants and clerks. They also increased the number of cloth types imported regularly.

This represents the initial phase of expansion. It is a phase which might last anything from two to 15 years. A business must first establish its profitability. It must, in other words, establish its competitive position within the particular economic niche in which it operates, build up collateral or generate enough risk capital to embark on the next stage of expansion. At this point the proprietor of the business must decide how to expand. The evidence suggests that he is unlikely to expand by doing more of the same. Rather, he expands vertically or differentially. He builds up a portfolio of small businesses within a single firm. Sometimes he sheds earlier profitable businesses in order to expand others. I have discussed this process as it occurs on a broad, communal scale, and as it relates to the growing availability of a family labour force.[21] In the following account I focus on the varying personal routes immigrant entrepreneurs adopt as they build up complex economic portfolios.

It became clear from the data that a business does not have to be very big in order to start diversifying. Even businesses with relatively small turnovers of under half a million pounds per annum can build up portfolios of disparate firms. The difference in scale runs through all their businesses. A big firm has an aggregation of big businesses, a small firm of small businesses. But both aggregate over time. This is shown in Table 5.1 which lists businesses with less than £2 million gross turnover, some considerably less.

The table lists the firms in descending order of gross turnover, based on approximate estimates in cases where firms are not limited companies or company records are unavailable or partial. (According to the 1985 Companies Act small firms may present modified accounts which do not include annual turnover. Many Asian firms utilize this exemption.)

One of the most important ways firms in the clothing and garment trade/industry diversify their portfolios is through investment in property, both commercial and residential (for letting). The clothing trade is a risky business with cyclical ups and downs and endemic cash-flow problems. Rather than continue expanding, and risk getting caught in a recession, manufacturers and wholesalers prefer to invest excess profits in property. A property generates a fixed, predictable income and also constitutes a hedge against inflation. Almost all successful businesses in Manchester have property companies kept separate from the main firm (which might be forced by an irate creditor into liquidation at any time).

Table 5.1 Small company portfolios (under 2 million annual turnover)

Firm (by size and type)	*Markets, retail*	*Manufacturing*	*Subcontracting*	*Wholesaling*	*Import/ Export*	*Property*	*Other*
1. (C)	(1983)	+	+	+	–	+	
2. (C)		+	+			?	trimming
3. (K)		+					
4. (C)				+		+	
5. (K)		+			–(P)	+	
6. (C)	+	+			–(P)		
7. (C)		+					
8. (C)				+			restaurants
9. (K)				+			
10. (C)	+		–	+	+	?	
11. (K)	+			+			
12. (C)		+	+	planning			
13. (C)		+					tried various
14. (C)*		+			+		construction
15. (K)		+					taxi-rank
16. (C)				+			
17. (C)		+					
18. (C)	+	+					was cinema owner
19. (C)	+			+			
20. (K)	+	+					

Key:
(C): Clothing
(K): Knitwear
(P): personal
* This firm is ranked by its clothing sales in Britain, not as an international firm. The main weight of the business is in construction.

Apart from commercial properties, Asians in Manchester have moved into the private rental property market in considerable numbers. Much of the student and bed-and-breakfast rental property (the latter subsidized or fully paid for by the state) is in the hands of Asians. Not all of them started in the clothing trade and indeed, apart from Asian property companies specializing in this field, there is also a good deal of small-scale investment by individual householders who may own only one rental property. In any case, for those in the garment and clothing trade/industry, property is considered a good back-up option for their business. Apart from the income it generates, properties provide collateral for short-term loans if and when a business encounters cash-flow problems.

Apart from property investments immigrant firms may aggregate a variety of different businesses together. Although even small firms often build up quite complex portfolios it is clear that the larger, more established firms tend to have more complex portfolios than recently established and smaller firms.[22]

Immigrant firms follow predictable patterns of vertical expansion – from market trading into manufacturing and/or wholesaling, from wholesaling into importing, from textile importing into conversion, etc. They also tend to expand horizontally in predictable ways: from wholesaling or manufacturing in one city to wholesaling or manufacturing in several cities (particularly London); from market retailing in several markets to High Street retailing in different locations. Manufacturers also expand into other related operations such as trimmings or stonewashing. Many knitwear firms with enough turnover do their own bulk importing of yarn. Almost all the firms invest in property, if only by buying their own premises. But each firm also tends to develop its own specific, often somewhat idiosyncratic, business portfolio in response to special opportunities or the availability of trusted expert personnel.[23]

A. Francis[24] has argued that 'marginal' entrepreneurs in Britain have tended to be motivated to achieve high rates of growth and profit from business because of their particular structural position within British society. In this they have differed from 'non-marginal' entrepreneurs who did not appear to have a very strong growth objective. He goes on to argue that 'To the extent that businesses founded by these non-marginals grew into today's large corporations, this appears to have happened by the defensive merger of a number of small businesses, rather than by fast organic growth'.[25]

The firms studied here have all tended towards high rates of growth. Yet it is not clear that developing a complex economic portfolio is often the best strategy for successful expansion. Some firms do better by reinvesting in the operation they are already familiar with. A case in point is that of one of the largest textile merchants in the city, who has a relatively simple portfolio. On the whole, however, the preference appears to be to develop a complex portfolio, most often in relation to the original business set-up. Developing a complex portfolio is a means of beating the competition through vertical supply lines and distributing risks (as in property investments) in a highly volatile and unpredictable trade. The evidence suggests that both methods – linear expansion or complex expansion – may be successful within a certain size-range. Beyond this a firm must develop a complex portfolio. In doing so, it must establish new channels of communication, different forms of expertise, a complex range of suppliers and customers and a complex division of labour. The development of economic portfolios thus often involves taking on new partners and hiring new managers. The test of a firm's success is its ability not only to create complex portfolios but to manage and control them.

Recognizing the tendency of immigrant entrepreneurs towards 'complex' rather than 'linear' expansion also has implications for the general study of ethnicity in industrial societies. Immigrant entrepreneurs utilize capital built up in industries requiring low levels of start-up expertise and capital investment, in order to expand into more sophisticated and highly capitalized industries. Their penetration of more complex economic fields is gradual and may occur over several generations, as expertise becomes available and management structures are built up. Their motivation towards continuous growth means, however, that in the long run they penetrate further and further into the economic life of their newly adopted country.

There is a further dimension to this growth and expansion revealed in the growth of a knitwear enclave in Manchester. Although a few Asian-run knitwear factories have existed in the city for almost 30 years, most knitwear was purchased by Asian wholesalers in Leicester. As Asians have penetrated into knitwear manufacturing in Leicester itself, new Asian knitwear factories were founded in Manchester as well. Indeed, some local manufacturers have shifted from clothing production to knitwear very successfully. This small knitwear enclave is gaining in sophistication, and most knitwear manufacturers have a range of customers, both local and national. Several of the more organized manufacturers supply the catalogue firms based in the North-West and Yorkshire.

Finally, the growth of the enclave has been characterized by a search for labour. Just as the Jewish enclave died when its natural supply of cheap labour disappeared, so too Asian manufacturers in the city are short of labour (the Asian population of Manchester city is relatively small and prosperous to boot). This has pushed the clothing trade further and further north, into Lancashire and Yorkshire: Rochdale, Bolton, Bury, Huddersfield, Blackburn and Nelson are all towns where Manchester manufacturers and wholesalers sub-contract to small, family-based local Asian (primarily Pakistani) firms. Links with immigrants in these northern towns have enabled some local manufacturers from Jhelum or Mirpur to build up their manufacturing base. Almost all the local clothing manufacturers I interviewed, however, whatever their regional origins, sub-contracted to firms in these towns. Similarly, the most successful of the retailers or manufacturers emerging in these outlying towns appear ultimately to relocate and open wholesale establishments in Manchester. The wholesaling centre thus includes successful Asians from beyond the city as well as local Mancunian entrepreneurs.

This leads to a more general point regarding ethnic entrepreneurship: immigrant entrepreneurship in the clothing/garment trade industries has always been, and has remained, a national phenomenon, characterized by its own division of labour and specific local-cum-ethnic monopolies and specialisms. London is still the centre of dress, ladies blouses and fashion manufacturing, with Greek and Turkish Cypriots apparently totally dominating this market. Rainwear and overcoat manufacturing, I was told, is dominated by Sikhs in Birmingham. Leicester dominates knitwear manufacturing. Manchester and the North-West is controlled mainly by Pakistanis and specializes in casual wear – trousers, skirts and co-ordinates, with some jeans-manufacturing as well.

ENTREPRENEURSHIP AND THE CONSTRUCTION OF KNOWLEDGE

Immigrant entrepreneurship in general, and in the clothing trade in particular, has been explained in terms of the low levels of start-up capital required to establish a business within enclaves such as the clothing trade. To some extent the whole discussion of entrepreneurship has focused on the issue of capital mobilization (through personal savings, loans between co-ethnics or rotating credit associations) and the availability of labour (familial and cheap). In other words, the apparently 'hard' economic features of ethnic entrepreneurial strategies have dominated sociological discussions. Other features of the trade, such as its reliance on flexible and quick responses or its individual-based type of mechanization are also essentially pragmatic explanations of entrepreneurial success. The further step of asking why certain enclaves require low capital investment at one level, and yet involve enormous capital investments (and profits) at another level, is not usually asked. One response would be to deconstruct the nature of knowledge in the fashion industry to discover what this can tell us (a) about the ability of immigrants to penetrate this field, and (b) the underlying values of interaction in the trade.

It is clear that ethnic firms tend to develop economic monopolies by clustering within enclaves. This would seem to point to a sharing of business information and experience. Interaction in the trade is dominated, however, contrary to expectations, by secrecy, even among co-ethnics. Competing manufacturers or wholesalers are extremely suspicious of one another and unwilling to share information, even at the rudimentary technical level. Business contacts are jealously guarded. Most businessmen are not allowed into manufacturing premises. My own requests to be allowed to observe businessmen in action were universally rejected, even by close friends. Yet it is clear that knowledge must be shared and that immigrants do learn from each other. Trying to explain this conundrum and the unwillingness of entrepreneurs to be observed as they conducted their businesses, became a major objective of my study. Let me stress from the start that I believe that illegal practices, which I deal with below, are only a very small part of the explanation of this phenomenon. Secrecy is the hallmark of all business transactions, even the most legal. To understand the prevalent secrecy and suspicion in the trade, I tried to discuss the kind of information businessmen keep secret with the various entrepreneurs I interviewed, with significant results.

Academics tend to think of knowledge as publicly available information. A further evolutionary assumption is that modern industrial production is highly integrated assembly-line production, requiring accurate, standardized and holistic systems of knowledge, and hence also requiring a regimentation of the labour force and a very complex division of labour, also associated with systematic modes of control and surveillance. Such integrated knowledge systems may indeed possibly be required in the manufacturing of complex machinery such as cars or aeroplanes. They are not, however, a necessary feature of clothing trading and manufacturing. Here knowledge can be broken down, it is partible. Moreover, a great deal of the knowledge required is a knowledge of tiny details, extremely difficult to teach, what we might call 'experience' or 'practical knowledge' in the sense used by Bourdieu. It is a knowledge gained through trial and error, an artisan's type of knowledge. It is also a sense of timing, of knowing when to buy, how to give or get credit, how to make deals, how to build up trust. Such knowledge is acquired at great personal cost, rather than being publicly available. Further, knowledge in the fashion industry, as in all industries, is partly innovative and creative; it is changing knowledge, which requires a mixture of experience and imagination. The fashion industry has its own industrial secrets – new fabrics, the forthcoming spring colours, the length of skirt, etc. All this experience and creativity is, however, useless without 'knowing' the right customers and suppliers. Building up a business network is also a form of knowledge which involves time and personal effort, and it is rarely a pre-given.

For small firms in the trade knowledge is thus partible, limited, and regarded as personal property. Because it is personal it is secret, not to be shared. By contrast, for the large firms in the trade, only innovation, 'real' fashion secrets, are regarded as personal property. Other forms of knowledge such as technical expertise can be purchased as a commodity on the open market. The main achievement of the large firms is the creation of systems of production and quality control. The trade thus encompasses a continuum from an extreme where all knowledge is personal property and thus secret, yet at the same time being fundamentally imperfect and limited in scope, to knowledge which is integrated, commoditized and systematic, and secret only in its innovative aspects. This continuum can be presented as follows:

KNOWLEDGE

Personal property, secret (acquired through hard work)	Public commodity
	Fragmented, partible (limited expertise of single production line)
	Integrated, holistic (total and coordinated expertise of many production lines)
Innovative, creative, uncertain (fashion; industrial secrets)	Technical, quality control ('finish')

The third and final opposition cuts across the other two. The movement is towards greater technical expertise and greater innovation, at the same time that there is also a move towards purchasing expertise as a commodity, and a more highly integrated and systematic control of production lines.

The fact, however, that the total process of production and sale can be broken down into constituent parts, and that quality control, while determining the price, is not essential, is the reason the trade is open to inexperienced newcomers. An immigrant may specialize in making Terylene trousers and nothing else. He may make them badly and yet be able to sell them if the price is low enough. He may have only one business customer or supplier when he starts. He gains experience slowly and painfully, and considers this experience to be his own personal property. This makes it clear why he will not answer questions put by researchers: as many manufacturers or traders told me: 'I spent three weeks finding that out, and you want me to give you this information just like that!'

Knowledge is thus subjectified and personalized. The theme most reiterated by entrepreneurs is that their success has been due entirely to 'hard work'. None mentioned any intrinsic intellectual advantage they had over others – for those without training or knowledge the effort involved in gaining knowledge, measured both in time and in financial failures, can only be summed up in the euphemistic notion of 'hard work'.

The building-up of technical and innovative expertise goes alongside the building-up of network contacts. Information in the clothing trade passes vertically, from retailer to wholesaler to manufacturer to fabric wholesaler. They keep each other informed on the 'lines' that are currently doing well. Some wholesalers retain a market stall just in order to keep in touch with changing tastes. Secrecy and trust can thus be regarded as opposed characteristics of interaction. An entrepreneur builds up trust with suppliers and customers and this trust involves sharing knowledge and information; he keeps his trusted contacts, as well as the knowledge he has acquired, a secret from everyone else. Absolute trust is, however, rare; most traders and manufacturers have secrets they share with no one outside the firm itself.

Moreover, the secretiveness between manufacturers or retailers can be manipulated by large wholesalers and used to undercut prices. Attempts by manufacturers to set up co-ordinating associations have mostly failed, because struggling manufacturers suspect that their more successful colleagues wish to 'discover' who their customers are and 'steal' them. It is only the more confident established manufacturers who are willing to share knowledge. This has to do partly with the kind of knowledge needed to ensure financial growth. Although fashion secrets may seem to be the most critical here, in reality technical expertise and systems control are more important for financial growth and expansion. A large firm like Marks and Spencer has complex control systems which ensure that the 'finish' of their clothing retains its high reputation. It is noteworthy, however, that the firm does not need to retain production entirely in its premises–most manufacturing for the big High Street retailers is done on a sub-contracting basis to smaller firms. What such firms have is a highly efficient and integrated system of production and supervision, contrasting markedly with that of immigrants at the cheap end of the trade. However, as immigrants gain technical expertise, they begin to control quality, at both the manufacturing and the retailing level. So the movement is towards more systematic forms of knowledge and hence of production. As they begin to build up complex portfolios, they also purchase more expertise as a commodity on the open market: they hire production managers, designers, salesmen, etc. They evolve management and communication systems within the firm. Hence capital accumulation and the accumulation of knowledge parallel each other and are intrinsically related. Profits (as well as turnover) ultimately depend on the level of integrated expertise a firm develops. As expertise becomes more widespread some aspects of this knowledge become less personalized, in the sense that entrepreneurs take it for granted. They no longer regard it as 'secret', the product of 'hard work'. This knowledge comes to be part of a collective pool which can be tapped by aspiring immigrant newcomers to the trade.

THE DEVELOPMENT OF INSTITUTIONAL RACISM

A further reason why immigrants do not wish to be observed, or their affairs pried into, is linked to the level of harassment directed against them. Like most forms of institutional racism this harassment may have some objective basis to it, but its effect on the trade has been almost entirely negative. The harassment comes from social security officers attempting to discover workers in the trade who are claiming social benefits, and from tax inspectors attempting to discover tax evasions. There have been constant raids on factories, most of them inconclusive but causing enormous resentment and disrupting work.

I did not attempt to investigate illegal practices in the trade, but my guess is that these are far less significant than is commonly thought in determining the survival and growth of the enclave. The fact, however, that there were tax and social security 'spies' infiltrating the trade made my own research more difficult. It was also evident that they created an 'us/them' confrontational atmosphere which exacerbated the sense of isolation Asians, and particularly Pakistanis, feel. Given the incredibly hard work and long hours Asians put into their businesses, this harassment is simply regarded by most of them as an attempt to prevent the community from achieving economic prosperity.

A further and very serious form of institutional racism has also developed in relation to insurance company practices.[26] These have been inspired by a run of insurance claims made by Asian businessmen for fire damage, which were investigated and found, in some cases, to be caused by arson. As a result, insurance companies now demand exorbitant annual premiums from all traders and manufacturers and in most cases refuse to insure altogether. Even long-term clients with an entirely clean record, who have never claimed insurance, have their premiums doubled and tripled quite arbitrarily. The situation is so severe that many of the smaller firms have no insurance at all. This can precipitate extremely serious situations as traders and manufacturers see their life's work vandalized or burnt, without any financial recourse. There have been cases in Manchester of whole mills burning down, with as many as ten factories, none of which were insured. It would be absurd to suspect arson in such cases. The municipal corporation's paltry attempt to set up an insurance scheme has had no impact on the problem. Much of the damage to premises is caused by vandalism, often racially motivated, which the police is unable to control. Hence the syndrome of government harassment, inadequate protection and inadequate insurance creates a complex of disadvantage for Asian immigrant entrepreneurs which goes almost entirely unrecognized. Immigrant organizations have attempted to deal with this problem, and sometimes co-ordinate the hiring of security services, but without adequate insurance back-up no firm is safe. The sense of alienation from the wider society that these multiple forms of harassment imply is increasing and should be addressed by policy-makers.

Asian entrepreneurship in Manchester has thus been associated with two contradictory trends. On the one hand, the entry into business has involved a major process of learning and integration into the wider society. Moreover, as traders or manufacturers prosper, they move beyond the inner city into primarily non-Asian suburbs, further accelerating their sense of integration into British society.[27] On the other hand, however, secretiveness in the trade, suspicion of state agents, resistance to any outside interference and problems of inadequate insurance tend, seen together, to intensify processes of social encapsulation, alienation and ethnic cultural-cum-religious enclavement. Which of these opposing processes will dominate is historically determined. The tension between the two trends remains unresolved. The capturing by sojourning migrants of economic enclaves, together with the economic resources these generate, cannot be regarded as leading simply to an irrevocable process of ethnic polarization.[28] While certain historical conditions may precipitate such polarization, counter-tendencies towards accommodation and integration may often be equally powerful. Both processes may, indeed, intensify simultaneously, but the more open, public and accountable businesses become, the less likely their owners are to tolerate calls towards extreme ethnic exclusivism.

CONCLUSION

Post-war Manchester has been a city in industrial and commercial decline. Going against the trend, and disregarding, apparently, the declining opportunity structure, South Asian entrepreneurs, predominantly Muslims, have successfully entered and revived some parts of the clothing industry, creating once again a thriving manufacturing-cum-trading enclave. They have also created an expanding textile trading niche.

Inter-ethnic competition, succession and complementarity within the textile and clothing trades, dominated first by Jewish entrepreneurs and more recently by Asian entrepreneurs, reveals certain regularities significant for a comparative analysis of immigrant entrepreneurship: First, the structure of immigrant enclaves may, and indeed, has, remained relatively constant, irrespective of the ethnic group monopolizing the enclave. Second, succession does not necessarily entail conflict, although recent immigrants of different ethnic origins compete with one another and may ultimately establish a perceived hierarchy or hegemonic structure of relations between the different groups and subgroups.

The success of immigrant entrepreneurs as traders and manufacturers which is evident in Manchester's history of immigrant succession points to the fact that immigrant entrepreneurship must be understood as a cumulative process taking place over time, within specific historical contexts.

Following an earlier discussion of collective processes of niche expansion, I argued here that immigrant entrepreneurs – and the economic enclaves they monopolize – grow through both linear and complex expansion. Moreover, to understand the symbolic value attributed to work and secrecy in such enclaves, the need is to analyse the nature of knowledge in the industry.

Recognizing the centrality of knowledge and secrecy is crucial for an understanding of both how immigrant entrepreneurs behave and the basis for immigrant concentrations in certain types of industry. Hence, the previous exclusive focus on the economic and pragmatic features of industry or trade, is theoretically inadequate. The need is to recognize the unique and complex nature of knowledge as an integral sociological aspect of any particular industry. To do so, the very concept of knowledge has to be deconstructed. Knowledge is not a unitary concept, or a constant. This is critical to an understanding of its determination of structures of trading and manufacturing in general, and of immigrant entrepreneurship in particular. The complex construction of knowledge that immigrant entrepreneurs hold is made manifest in the extreme secretiveness characterizing relations within the trade, even between members of a single ethnic group.

This secretiveness also provides a clue to the increasing institutional racism prevalent in the trade, perpetrated mainly by government offices and private companies against immigrant businesses. Both these dimensions – the nature of knowledge and the increase in institutional racism – are critical, pervasive features of immigrant entrepreneurship.

REFERENCES

1. See E. Bonacich, 'A Theory of Middleman Minorities', *American Sociological Review*, 38 (1973); E. Bonacich and J. Modall, *The Economic Basis of Ethnic Solidarity: Small Business in the Japanese American Community*, (Berkeley: University of California Press, 1980); E. Bonacich and I. Light, *Immigrant Entrepreneurs: Koreans in Los Angeles 1965–1982* (Berkeley: University of California Press, 1988); I. Light, *Ethnic Enterprise in America: Business and Welfare among Chinese, Japanese and Blacks* (Berkeley: University of California Press, 1972); I. Light, 'Immigrant and Ethnic Enterprise in North America', *Ethnic and Racial Studies*, 17 (1984), pp. 195–216. K.L. Wilson and W.A. Martin, 'Ethnic Enclaves: A Comparison of Cuban and Black Economies in Miami', *American Journal of Sociology*, 88 (1982), pp. 135–60; A Portes and R. Bach, *Latin Journey: Cuban and Mexican Americans in the United States* (Berkeley: University of California Press, 1985); R. Ward and R. Jenkins (eds), *Ethnic Communities in Business* (Cambridge: CUP, 1984).
2. Bonacich, 'A Theory of Middleman Minorities'.
3. Light, *Ethnic Enterprise in America*, and Light, 'Immigrant and Ethnic Enterprise'; Bonacich and Light, *Immigrant Entrepreneurs*, pp. 178–204.
4. Bonacich, 'A Theory of Middleman Minorities.'
5. Light, 'Immigrant and Ethnic Enterprise'; Portes and Bach, *Latin Journey*.
6. H. Auster and H. Aldrich, 'Small Business Vulnerability, Ethnic enclaves and Ethnic Enterprise', in Ward and Jenkins (eds), op. cit.
7. A. Phizaklea, 'Entrepreneurship, Ethnicity and Gender', in S. Wentwood and P. Bhachu (eds) *Enterprising Women: Ethnicity, Economy and Gender*, (London: Routledge, 1988).
8. F. Barth, *The Role of the Entrepreneur in Social Change in Northern Norway* (Bergen: Scandinavian University Books, 1963); R.A. Cohen, *Custom and Politics in Urban Africa* (London: Routledge & Kegan Paul, 1969); U. Hannerz, 'Ethnicity and Opportunity in Urban America', in R.A. Cohen (ed.), *Urban Ethnicity* (London: ASA Monographs 12, 1974), pp. 37–76.
9. P. Werbner, 'Business on Trust: Pakistani Entrepreneurship in the Manchester Garment Trade', in Ward and Jenkins (eds), op. cit., pp. 166–88; Werbner, 'Enclave Economies and Family Firms: Pakistani Traders in a British City', in J. Eades (ed.), *Migration, Labour and the Social Order* (London: ASA Monograph 25, 1987); Werbner, *The Migration Process: Capital, Gifts and Offerings among Overseas Pakistanis*, (Oxford: Berg, 1990).
10. R. Mellor, 'Manchester Inner City', in H. Newby et al. (eds), *Restructuring Capital* (London: Macmillan, 1985).
11. Mellor, op. cit., pp. 4 and 15.
12. P. Werbner, *The Migration Process*.
13. Cf. R. Ward, 'Ethnic Entrepreneurs in Britain and Europe,' in R. Scase and R. Goffee (eds), *Small Business in Europe* (London: Croom Helm, 1986).
14. For an extensive discussion, see Werbner, *The Migration Process*.
15. S. Nowikowski, 'Snakes and Ladders: Asian Business in Britain', in Ward and Jenkins (eds), op. cit.
16. Werbner, *The Migration Process*, ch. 1.
17. Werbner, in Eades (ed.), *Migration, Labour and the Social Order*.

18. Werbner in Ward and Jenkins (eds) op. cit.; and *The Migration Process*.
19. Werbner in Ward and Jenkins (eds), op. cit.; Werbner, in Eades (ed.), op. cit.; Werbner, *The Migration Process*.
20. Werbner, in Eades (ed.), op. cit.; Werbner, *The Migration Process*.
21. Werbner, in Eades (ed.), op. cit.
22. See P. Werbner, 'Building Business Portfolio: Immigrant Entrepreneurs and Small Firm Expansion' (1988 Conference Paper: n.p.).
23. On this feature of multiple enterprise, see N. Long. 'Multiple Enterprise in the Central Highlands of Peru', in S.M. Greenfield, and A. Strickon and R.T. Aubey (eds), *Entrepreneurs in Cultural Context* (Albuquerque: University of New Mexico Press, 1979).
24. A. Francis, 'Families, Firms and Finance Capital: The Development of U.K. Industrial Firms with Particular Reference to their Ownership and Control', *Sociology*, 14, 1 (1980), pp. 1–28.
25. Ibid, p. 6.
26. See also S. Patel, 'Insurance and Ethnic Business – An Emerging Crisis in the Inner City', *New Community*, 15, 1 (1988), pp. 79–90.
27. Werbner, *The Migration Process*, ch. 1.
28. Bonacich, 'A Theory of Middleman Minorities'.

6 Why Move? Regional and Long Distance Migrations of Gujarati Jains

Marcus Banks

INTRODUCTION

The anthropologist studying migration is in the position that he or she must consider it *ex post facto*, or at least as a process that has already been set in motion (as indeed must be done with many other facets of social life). All explanations as to the cause of migration are therefore made with the benefit of hindsight and carry with them a sense of inevitability. Far more difficult then, perhaps even more difficult than predicting which sections of populations may migrate in the future, is to explain why certain populations have not migrated. While the explanation may cause difficulty, it should not – in theory – be too difficult a problem to investigate in the field. A difficulty arises, however, when a broader historical and geographical context is taken into account, as I shall discuss.

Similarly, when anthropologists (and other social scientists) consider migration they tend to make two assumptions. First, that 'voluntary' migrations, that is, not the forced migrations of refugees, for example, are primarily labour migrations. Secondly, that the migrants form a group in some sociological sense. The assumption that a migration of people is essentially a migration of labour is often motivated by the observer's interests; if the observer is an economist, an economic anthropologist, or an economic historian, then labour and its distribution are the prime object of study and the movement of people thus becomes symptomatic rather than causal (Adam Smith's 'vent for surplus'). Janet Abu-Lughod likens the human subjects of such accounts to iron filings, drawn irresistibly by forces beyond their control.[1] Even when the investigator's primary interests are not economic, it is not difficult to find some material or economic benefit which results – or was intended to have resulted – from a voluntary or indeed a forced migration. One might cite as a case in point the forced repatriation of Vietnamese 'boat people' from Hong Kong, on the grounds that they were 'merely' economic refugees. Certainly migrants themselves may cite economic factors as their motivation for moving. The only alternative to this political economy approach is a weakly psychological view, which while refining or rejecting earlier 'bright lights' theories of motivation[2] nonetheless eschews macro-economic causality in favour of individual choice and micro-sociological context.[3] While a stress on individualistic motivation may, with sensitive ethnography, provide a richer, more humanistic account,[4] it often does so at the expense of a consideration of wider socio-economic factors which impinge upon the migrant.

The second assumption – that migrants form a group – is similarly *a priori* although, as such, is not particularly unfounded. Anthropologists are used to dealing with groups and to ascribing individual choices and actions to manifestations or expressions of group desire or will or – more mystically – to some transcendent set of moral forces (the Durkheimian inheritance of anthropology). The danger lies again in seeing migration as symptomatic of group behaviour rather than as independent of it or even as causal of it.

When one considers migration in Asia, and particularly within and from India, a number of further anthropological assumptions come to the fore. India, perhaps more than any other nation-state in the world, is seen by anthropologists, and many others, as being divided into absolutely discrete population units, these units being unique cultural islands. There are several of these units, the most basic perhaps being the so-called castes. Other, larger scale units, are villages, religious groups, linguistic groups, and regional units, which are often roughly coterminous with linguistic groups. The work of colonial administrators and scholars, followed by anthropologists and now indigenous and foreign journalists has served to bring into being and reify these units as conceptual categories that 'explain' social phenomena inside India – phenomena such as 'caste wars', 'reservation riots', 'communalism', 'Sikh terrorism', 'Kashmiri separatism' and such like. The mutual antagonism between the units presupposed by these explanations only serves to reinforce further their absolute boundedness. Thus it is no surprise that migration within and from India should be thought to be dependent upon and expressive of these units. For example, when sociologists and anthropologists discuss the 'ethnic minority' population of Britain the 'natural' units isolated are often regional-linguistic ('Punjabi community', 'Gujarati community') even though there may in fact be some confusion or overlap – between Punjabis and Pakistanis, for example. Such confusion may be resolved by adding a further order of classification – between for example, Punjabi Muslims and Punjabi Sikhs.

This chapter discusses the migratory strategies and outcomes of relatively small numbers of North Indians over the past 450 years. I wish to argue that these migrations form a set; to consider only one part of these migrations, as anthropologists and sociologists have tended to do – that is, to look at a particular 'journey' from, say India to East Africa – is to isolate a handful of data and subject them to a misleading interpretation. The migrations I am dealing with concern groups of Jains moving from Rajasthan down into Gujarat and then overseas to East Africa and the UK.[5] These 'groups', however, should not be taken as having any *a priori* existence; rather, I am suggesting that there is a dialectical process whereby migratory strategies serve to form and reify 'groups', while at the same time, the structure of such 'groups' influences the form and nature of the migrations. Indeed, there is an analytical point to consider here, which is the extent to which we can meaningfully talk of 'groups' at all. In this instance, within a bounded field of process (the migratory process[6]) which *may* be largely defined by economic criteria (for the analyst) there are myriad sub-processes each directed by the agency of certain individuals or complex agents (families, clans, etc.). The question – following the anthropologist Edmund Leach – is whether the transcendent features which appear to give unity to the whole bounded field of process are causative (for example, 'vent for surplus') or merely epiphenomenal, and that social phenomena have no unity but for their statistical agglomeration. To quote Leach talking about the 'groups' (kinship groups in this case) that were supposed to be important in the system of land tenure he investigated in Sri Lanka: 'The constraint imposed upon the individual is merely one of patterning and limitation... The group itself need have no rules; it may simply be a collection of individuals who derive their livelihood from a piece of territory laid out in a particular way.'[7]

As an adjunct to the argument that groups of migratory individuals need share no common, transcendent purpose, motivation or drive, I also wish to argue that a common popular and anthropological assumption – that Indian or Hindu society is divided into a number of discrete 'castes' – is similarly misfounded. The processes I have observed and describe in this chapter lead me to think that migration may be formative of castes – or better, may be a contributory feature in ideological processes of caste fission and fusion which are manifest in group behaviour as a result of constraining features of the social and even natural environment. In adopting this position I am clearly on the side of Dumont[8] in seeing caste as a relational phenomenon, although I would reject his extreme ideological determinism in favour of an historically and politically sensitive approach as that recently proposed by D. Quigley.[9]

OSWAL AND SRIMALI

Let me turn now from these abstract and rather pedantic introductory remarks to the actual data. I am primarily concerned with two groups of

Jains, known currently as Oswals and Srimalis. These groups are described in Colonial census and survey literature as 'castes' and members of them can be found spread widely across the Gujarat/Rajasthan area. Perhaps Richard Fox's term 'supra-local caste cluster'[10] better describes them (see Table 6.1).

Table 6.1 Oswal and Srimali

'Supra-regional caste cluster' *('caste')*	*Jati/nat* *('sub-caste')*
Oswal	Jamnagar Visa Oswal Halari Visa Oswal (other Oswal-Kacchi, etc)
Srimali	(Jamnagar) Visa Srimali (other Srimali)

As castes, they fall into the middle range of the supposedly normative caste hierarchy and their members are 'traditionally' associated (both by themselves and by observers) with trade, money-lending, shopkeeping and other mercantile activity. My focus is narrower than this, however, and lies with three particular subsections of these two castes: two of the Oswal caste: the Halari Visa Oswals and the Jamnagari Visa Oswals; and one of the Srimali caste: the Jamnagar Visa Srimalis. The older anthropological literature on India would have described these as 'sub-castes' or 'local' castes; elsewhere I have used the term *jati* and this, or *nat*, is how members would describe such a grouping. The important point is that these are territorially limited and largely endogamous groups. That is, there are several groups of people calling themselves Oswal (for example) throughout contemporary Gujarat and Rajasthan, but each group will have one or more descriptive terms (possibly territorial) prefixing the name Oswal, will not exchange marriage partners with other Oswal groups and may deny that there is any link or common origin which unites all such Oswal groups. Members of all three *jatis* are – with a tiny number of exceptions – Jain by faith (mixed Jain and Hindu *jatis* are found elsewhere). I should stress now that I believe that Jainism has little or nothing to do with the migrations I am going to describe. I focused my attention in the field on Jain *jatis* for other reasons and, because of my knowledge of these three *jatis*, it is these that I am using as examples; the various processes and strategies I describe, however, ought to be applicable to other trading *jatis* in similar structural positions. I should perhaps also stress that in what follows I am implicitly referring to Jain men and to the migrations of them and the households they headed. The role of women in these migrations, as in so many others, has yet to be explored,[11] although it is clear that certain gender differentials were at work (for example, in the bringing over of brides for men born in East Africa). In the early 1980s I conducted fieldwork in the English city of Leicester and in the Saurashtran city of Jamnagar with Visa Srimalis, Halari and Jamnagari Visa Oswals. The total populations for each group at that time were approximately as shown in Table 6.2.

Table 6.2 Selected Jain populations in India, East Africa and England

Jamnagar (1980s)	Visa Srimalis	[± 2000]
	Halari Visa Oswals	4500
	Jamnagar Visa Oswals	1500
Bombay (1980s)	Visa Srimalis	[± 1000]
	Halari Visa Oswals	15000
	Jamnagar Visa Oswals	2000
East Africa (1950s)	Visa Srimalis	± 1400
	Halari Visa Oswals	± 6000
Leicester (1980s)	Visa Srimalis	500
	Halari Visa Oswals	400

Note: All figures are approximations, but those in square brackets represent no more than an educated guess.
Source: author's research.

The migratory movements I will describe are fairly complex and span at least 400 years. Table 6.3 and Maps 6.1 and 6.2 therefore provide a quick synopsis and overview.

Table 6.3 Synopsis of Jain migrations

Migration	*Century*	*Cause*
Rajasthan to Kaccha	?	?
Kaccha to Jamnagar	16th	'invitation'
Jamnagar to Bombay	19th	entrepreneurial expansion
Jamnagar to East Africa	19/20th	entrepreneurial expansion/ entrepreneurial entrée
East Africa to UK	mid-20th	forced/political
East Africa to North America	mid-20th	forced/political/ entrepreneurial expansion

Source: author's research.

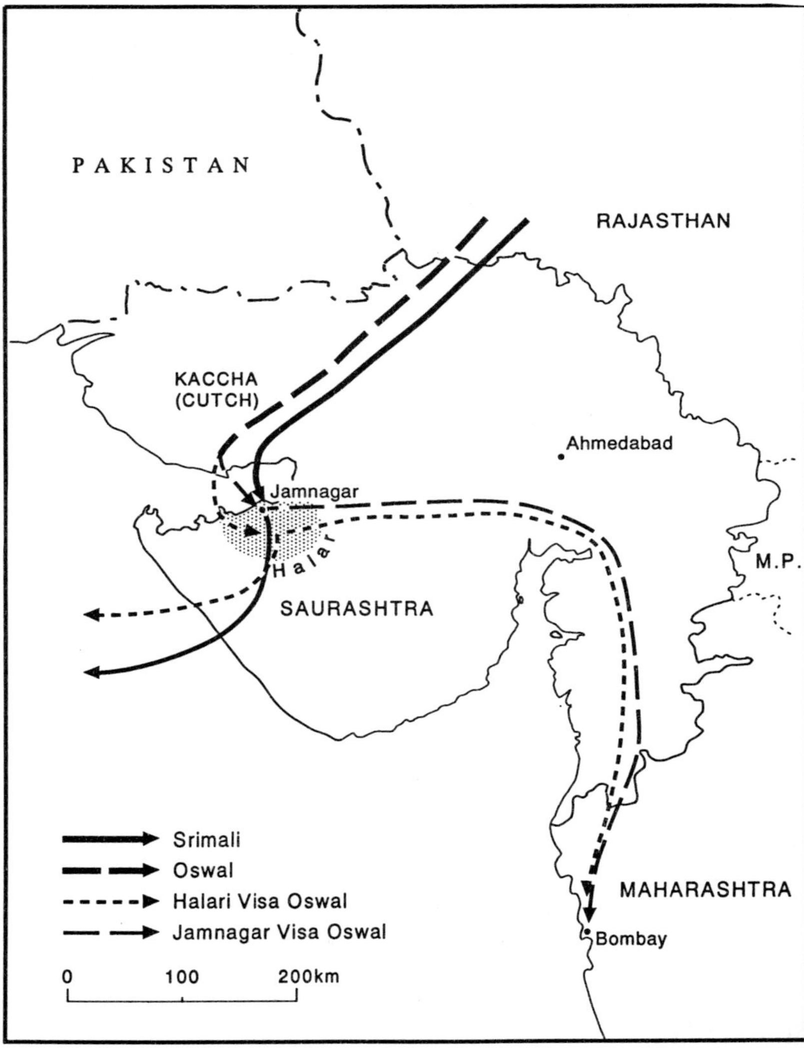

Map 6.1 Srimali and Oswal migrations in western India

RAJASTHAN TO SAURASHTRA

Earlier literature on the wider case groupings of Oswal and Srimali states that members of the two groups claim to have come originally from Rajasthan and to be of Rajput descent.[12] While no one I interviewed in Jamnagar or in

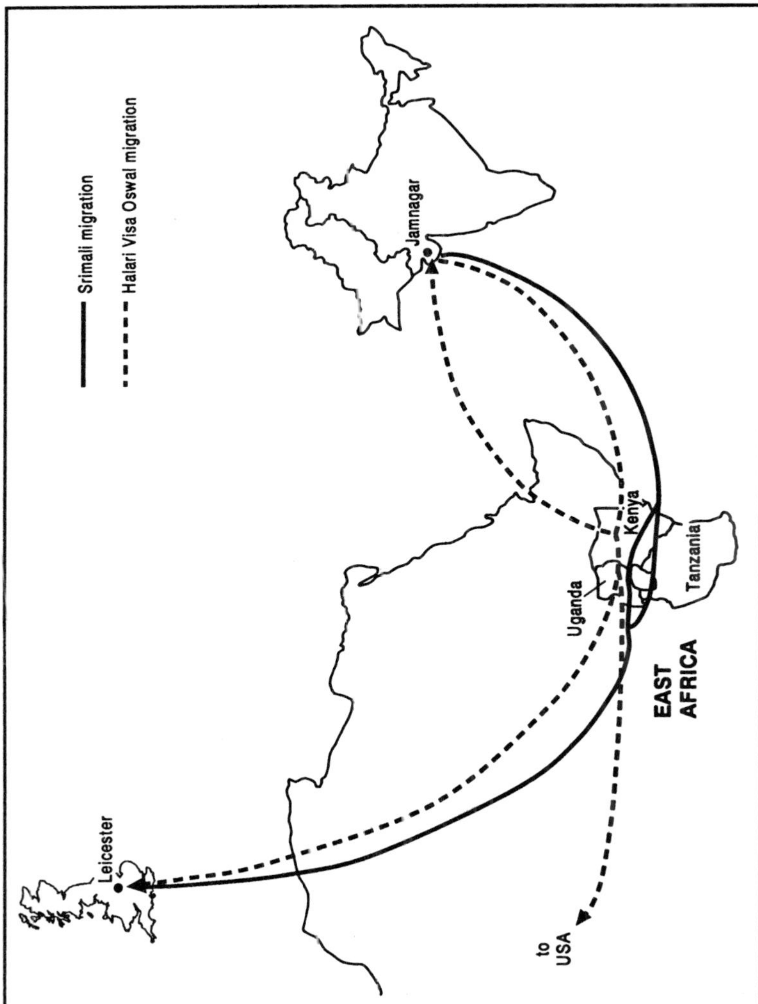

Map 6.2 Srimali and Halari Visa Oswal migration routes

Leicester in the early 1980s volunteered this information or could either confirm nor deny it when I asked, the existence of the claim is unsurprising. Many groups in India claim Rajput descent (the President of one of the largest Jain trusts in India, an Oswal from Ahmedabad, told me the Oswals were originally Hindu Sisodia Rajputs who 'converted' some 400 years earlier) and there are certainly many powerful and influential Jain traders to be found in Rajasthan today, as well as in the past.[13]

So the story of the Jamnagar Jains' migrations begins at some unknown point in the past, and some 300 miles distant from Jamnagar. There is a mythical quality to this starting point, receding as it does beyond lived historical time and beyond a concrete experience of geography. This mythical quality is reinforced by the fact that some of the nineteenth-century literature cites unknown sources which claim that the Srimalis in particular were created by the goddess Lakshmi from her holy garland (*sri mala*).[14] It is a long way from being a petal on a goddess's garland to queuing in the DSS in Leicester.

Assuming for the moment that there is something in the idea of mercantile Jains moving out from Rajasthan it must be recognized that this is a widespread diaspora, one that continues today with Rajasthani Marwari Jain traders to be found in all parts of India, as well as overseas, and that consideration of this migration has already been circumscribed by my own territorial interests.

Let us assume, then, that some of these Rajasthani Jains made their way across today's Gujarat state border into the desert region of Kaccha. Certainly it is from here that the Jains of Jamnagar (and thus Leicester) appear to have come and it is here that the mythical narrative of the migration becomes a historical one: we have a definite geographical location and a point in time somewhere at the beginning of the sixteenth century. At this time, much of Kaccha was under the control of Jadeja Rajputs who derived revenue for their kingdoms largely from trade – land trade with Central Asia and sea trade down the western coast of india. Agriculture would have been of secondary importance in such a dry area. D.P. Khakhar, writing in the *Indian Antiquary* in 1876, notes that the Kacchi Oswals are cultivators, while the Kacchi Srimalis are engaged in cultivation and trade.[15] However, while he implies that the Oswals and some Srimalis were actually tilling the land themselves, I would suspect that some of the Srimalis at least were actually landlords, employing sharecroppers or tenant farmers. I would also suspect that some at least of the Oswals were engaged in trade.

Some time in the 1530s one of the many disputes between rival factions of the ruling Jadeja Rajputs resulted in the leader of what was claimed to be the ruling lineage, Jam Rawal, being driven across the Gulf of Kuccha into

Saurashtra (the far western part of what is today the state of Gujarat) by an alliance of Emperor Humayun's Mogul army and some disenfranchised cousins. C.A. Kincaid, an English lawyer serving in British India, writing in 1931 but drawing upon what I take to be a dynastic chronicle says: 'When Jam Rawal's subjects learnt that he was about to leave them, not only his soldiers but a multitude of peaceful traders also resolved to go with their prince rather than be left to the tender mercies of a Moghul army.'[16] Jam Rawal, after a number of adventures, finally established a settlement in Saurasthra and called it Jamnagar. In some ways it is strange that the chronicle upon which Kincaid draws should cite such an inauspicious beginning for the foundation of Jamnagar: certainly the District Gazetteers (for example, that of 1884) point out that it was Humayun's pacification of Saurashtra that allowed Jam Rawal the chance to establish himself there. The 'peaceful traders' who loyally followed Jam Rawal seem to have included both Oswals and Srimalis, for within 24 years of Jamnagar's foundation (in 1540) the Oswals had built a Jain temple, to be followed 13 years later by the Srimalis.

Over four centuries later, Oswals and Srimalis I interviewed in Jamnagar were aware of their link with Jam Rawal, but phrased it in slightly different terms. They had, they claimed, been 'invited' by him to establish commerce and trade in the city, and to develop agriculture in the rural hinterland: many people in the city today seem to see Jam Rawal as a heroic pioneer, striding out bravely from Kaccha to conquer new lands. Again, this version of events, while not necessarily accurate, does have a ring of truth about it. Richard Fox claims that: 'In India, much of the history of town formation has to do with a local *raja*, *nawab* or big *zamindar* who established a town or created a market. This sort of petty nobility or local strong men endowed towns and began market places and, traditionally, provided supra-caste integration to them.'[17] In his discussion of 'Tezibazar', a market town he studied in Uttar Pradesh, he says: 'The zamindar attracted merchants to his market place by guaranteeing them protection – by overseeing the "peace of the market". He provided social control and organization for his artifically created urban locale.'[18] A further, and not necessarily exclusive reason for the merchants' arrival in Jamnagar is that of simple economic opportunism. A local Jamnagar author of this century, again drawing on what appear to be caste or dynastic histories, tells the story of Tejsi Shah 'of the Oshwal *jnati*', who came to Jamnagar in 1564 'for trade' and went on to build the first temple.[19] Similarly, Howard Spodek notes that some cities in Saurashtra, ruled by rulers of 'greater power and imagination' attracted 'élite groups' whose power and status were gained not through control of land (which was not in short supply in

India during the pre-colonial period[20] and certainly not in this sparsely-populated region), but through control of capital and human resources.[21] There is no doubt that Jains in the region generally constituted such an élite group and that, lacking fixed ties to the land in the main, were far more mobile as and when necessary: 'the city, and even the countryside, were not so much "home" as a base of operation'.[22] The picture so far presented then is that the Jains of Jamnagar arrived there either by flight, by invitation or simply for reasons of economic opportunism, and there they stayed for the next 300 years or so, with no significant further migrations, despite the apparent capacity to move on if necessary.

WHO'S WHO IN SAURASHTRA

The next phase of Jain migration begins in the mid- to late-nineteenth century, when Oswal Jains began to move to Bombay, and both Oswals and Srimalis started to journey overseas to East Africa. But there is a problem, for in the intervening three centuries the 'Oswals' and 'Srimalis' who came from Kaccha had become the 'Halari Visa Oswals', the 'Jamnagar Visa Oswals' and the '(Jamnagar) Visa Srimalis'. Where had these names come from, and why had the Oswals apparently split into two groups? The name question is the easier: the 'Visa' element refers to the supposed purity of descent of the group from a putative ancestor, a similarly named 'Dasa' group being a *jati* formed of the descendents of one who was once a member of the 'full' *jati* but fell from grace in one fashion or another. Such Visa and Dasa divisions are common throughout the region, especially among the mercantile *jatis*; in Jamnagar there are Dasa Srimalis (although I shall not deal with them here), but no Dasa Oswals to my knowledge.

The 'Halari' epithet for one of the Oswal groups is drawn from Halar, the name of the region immediately around Jamnagar city. It is clear that the apparently undifferentiated 'Oswals' from Kaccha divided into two groups some time after arriving in Saurashtra: those living in the city of Jamnagar (the Jamnagari Visa Oswals) and those who settled in small towns and villages nearby (the Halari Visa Oswals). Either that, or two distinct groups of Oswals from Kaccha migrated separately and somewhere in the next 300 years lost their original names and took on new ones.

In fact, the work of two anthropologists in the 1970s and 1980s seems to confirm that a single group split, rather than there being two groups in the first place. It is therefore the case that, in this context at least, migration and caste/*jati* organization should both be considered as processes, processes that bear significantly on each other.

Maureen Michaelson and J. Zarwan both worked with migrant Halari Visa Oswals, Michaelson in England and Zarwan in Kenya. The story they jointly tell is as follows. According to Zarwan, a split developed between those Oswals living in Jamnagar (who he refers to, rather confusingly, as 'Cutchis' – that is, from Kaccha) who were mostly traders, and those in the villages, who were agriculturalists.[23] Over time, the two groups, who had previously freely intermarried, became linked hypergamously: the village-dwelling Oswals (Halaris) giving daughters to the Jamnagar-dwelling Oswals. During a severe famine in the region (1899–1901), the Halari Oswals broke off these relations when, according to Michaelson, Oswals in Jamnagar were able to exploit the impoverished agriculturalists by purchasing brides from them. A.M.Shah notes that this is a common cause of endogamous unit fission in Gujarat.[24] The Halaris gained financially of course by acquiring bride-price instead of giving dowry, but in doing so lost immeasurably in status terms.[25] 'One can therefore date the final emergence of the Halari Visa Oswal community as an entirely separate caste from this time'.[26] Michaelson, working largely from Zarwan's material and information gathered from Halari Oswals now resident in Britain, agrees largely with this account but claims that in the mid-nineteenth century more Oswals came over from Kaccha ('known as Cutchi Visa Oswals or Jamnagar Visa Oswals')[27] and it was they who established the hypergamous relationship with the Halaris. On this point I think Michaelson has allowed herself to become confused by Zarwan – there are some Oswals living in Jamnagar known to Jamnagaris and Halaris as Kacchi Oswals, but they are few in number and continue to marry in Kaccha.[28] Those I spoke to saw themselves as temporarily resident in Jamnagar for reasons of trade.

In interviews I conducted with both Jamnagar-based Oswals (the Jamnagaris, as they refer to themselves) and the Halari Oswals who had settled in the surrounding villages, they denied any link or intermarriage between the two groups, and some even went so far as to deny any links whatsoever, claiming that the common term 'Oswal' in their *jati* names was a mere coincidence. (This may of course be merely a case of 'genealogical amnesia'.) The indigenous account is thus of separate migrations from Kaccha 400 years ago.

My opinion is that after an initial (possibly small) migration 400 years ago, in which an homogeneous Visa Oswal population settled in and around the town of Jamnagar, there has been a steady trickle of migration since, with the migrants going to stay – in the first instance – with relatives, wherever these were settled. Using the urban centre as a resource, those settled in the town entered into business and quickly sought to isolate themselves from their rural cousins, which also meant a partial denial of their Kacchi ancestry. (Most or many Halari Oswal male adults are able to speak the Kacchi dialect – which is largely incomprehensible to a Gujarati speaker – and were proud of this fact, whereas I met no Jamnagaris who admitted to this skill). New migrants from Kaccha were probably still accepted in the city, though as time when by the Jamnagar community would have grown sufficiently not to need these for marriage; they were probably expected to assimilate as quickly and thoroughly as possible, to preserve the stability of the business community.

In summary then, two groups of people who share a similar group name, today deny any common link and claim to have made separate migrations from the same place and for the same or similar purposes (one to develop trade, the other to develop agriculture). By the mid-twentieth century, members of the 'rural' group had also become involved in trade and are today to be found living and working in the city, alongside members of the city-based group.

SAURASHTRA TO EAST AFRICA AND BEYOND

The next significant phase of Jain migration I wish to consider began in the late nineteenth century and moves the chronology of the account on from historical time to remembered time or family history. Economic change within the British Empire generated a demand for skilled and entrepreneurial labour in the East African countries of Kenya, Tanzania and Uganda and there was a ready source of such labour in India. The Jamnagar Jains took advantage of this and began to migrate again. Simultaneously, the great famine of 1899–1901, while possibly having a decisive effect on the developing Halari/Jamnagari Oswal split, also provided an impetus for migration. Without crops, the agriculturalists of Halar (Oswals and many others), had no money to buy goods, nor could they provide the commodities that many Jamnagar traders (Srimalis and Jamnagari Oswals) depended on for trade.

Saurashtra in general experienced out-migration to East Africa, as did other parts of Gujarat, particularly Kaira District in Central Gujarat, the homeland of the Patidars or Patels. Of the three groups in this study, in and around Jamnagar (the Halari Visa Oswals, the Jamnagar Visa Oswals and the Visa Srimalis), two had fairly high rates of migration to East Africa from the end of the nineteenth century, while one did not – the Jamnagar Visa Oswals. This leads to a further problem – why didn't the Jamnagar Visa Oswals go? I can find no clear answer, except that of the three groups, the Jamnagar Visa Oswals are probably the most deep-rooted in the city of

Jamnagar and that, from some time in the mid-nineteenth century, they had begun to establish a '*jati* outpost' in Bombay: by the early 1980s there were more Jamnagar Visa Oswals in Bombay than in Jamnagar (around 350 households in Bombay as against 290 in Jamnagar), although there is frequent movement between the two cities and many Bombay-settled families maintain (unused) houses in Jamnagar. Certainly, the group seems to have chosen to expand its entrepreneurial activities within India, taking advantage of the British development of Bombay as the major port and commercial centre of western India.

By contrast, the Visa Srimalis and the Halari Visa Oswals can be said, at least for the purpose of argument, to have had a greater motivation for overseas migration or to be subject to predisposing factors. The Visa Srimalis are by no means as neatly defined an endogamous group as the Jamnagar Visa Oswals. They do not call themselves 'Jamnagar Visa Srimalis' (although there is some recognition of the rights families have in city-based corporate property, rights which depend on residence in the city over several generations[29] and individuals I have questioned were far less precise about which other Srimalis in the region they may or may not marry. In this sense they more closely mirror Spodek's loosely-tied élite groups[30] than the Jamnagar Visa Oswals. The difference between the Srimali and the Jamnagari Oswal *jatis* on this matter of endogamy or boundedness appears to be one of occupational strategy. For whatever reasons, Jamnagari Oswals had gained and retained very tight control of a limited number of enterprises in Jamnagar and Bombay, while the Srimalis – who in Jamnagar to some extent mix retail trade and commodity speculation with land ownership in the surrounding villages – have a more widely diffused field of economic strategies which would, I would argue, predispose them to take advantage of the new opportunities opened up in East Africa. While all Jamnagar Visa Oswal marriage alliances are today conducted exclusively between the seven hundred or so households in Bombay and Jamnagar, Visa Srimalis in Jamnagar will contract marriage with other Visa Srimali groups across a wide area: Christine Cottam cites evidence of Jamnagar Srimalis contracting marriage alliances with Srimalis as far away as Rajasthan.[31]

The Halari Visa Oswals, on the other hand, seem to have been seeking an entrée into commerce towards the end of the nineteenth century.[32] With a major drought across the region at the turn of the century, agriculture became less and less secure as an occupation, while simultaneously the British began to show an interest in developing Jamnagar as one of the model Native States of the region,[33] helping to stimulate commercial enterprise. As the poorer country cousins of a successful Jamnagar-based trading *jati*, the scope for Halari Visa Oswal enterprise in Jamnagar must have seemed limited, and thus they took advantage of the prospects East Africa offered. Thus the Halari Visa Oswal migration paralleled that of the (Jamnagar) Visa Srimalis in form, but was underpinned by rather different motivations and precipitating factors.

The migration to East Africa seems to have followed the usual lines of chain migration in the early years, with pioneer individuals making the journey and sending back to India for further labour, usually familial. Zarwan notes that, for Oswals at least, early 'push' factors were later replaced by 'pull' factors: 'Word would get around that conditions in Africa were good: other people would emigrate.'[34] The Halari Visa Oswal migration was exclusively from the *bavangami*, the 52 villages of Oswal residence around Jamnagar, although some migrants may also have spent a preliminary period in either Bombay or Jamnagar. (There is today a substantial population of Halari Visa Oswals in the Greater Bombay area, though much of this migration appears to have been since Indian independence). In contrast, the Srimalis in East Africa came from Jamnagar itself, and from smaller market towns in the region (generally to the east and south – Wankaner, Rajkot, Junaghad, Bhanvad) as well as from other towns and cities in Gujarat – Ahmedabad and Valsad for example. (More precisely, they did not come from Kaccha (Kutch), North or East Gujarat.) While the Halari Visa Oswals constituted an endogamous group and freely married within it (apart from rules of surname exogamy), the Srimalis in East Africa were less of an homogenous *jati* and even in 1982 Visa Srimalis of Saurashtran origin in Leicester (the majority) told me that they would not marry their children to Visa Srimalis of Ahmedabadi origin for preference. Although for many, migration to East Africa was permanent and preceded migration to Britain, there were many other options. Several Halari Oswals and Visa Srimalis had always planned to return to India, their timing perhaps prompted but not necessarily forced, by the East African political situation after Independence. In some cases, migration to East Africa could be surprisingly brief: the father of one of my Oswal informants spent only two years in Kenya (1935 to 1937) before returning to his village in Halar on the death of his father to take over the family farm. Similarly, many boys born in East Africa were sent to Jamnagar for their education, some returning afterwards and some not, while many women now in Leicester came to East Africa from India in adulthood, as brides.

Once in East Africa, Halari Oswals and Visa Srimalis followed different career and other strategies there. The Oswals, composed mostly of interlinked family groups, settled almost exclusively in Kenya and concerned themselves only with trade (retail and wholesale). The Srimalis, by contrast, were largely better educated (as agriculturalists the Oswals had seen little need for education prior to emigration) and, although some entered into trade, many others sought employment with the Colonial administration, particularly in Uganda and Tanzania. Up to 1940 there were only two or three Oswal households in what is now Tanzania, a number which increased to about twenty in as many years, the settlers all coming from Kenya. Uganda by contrast never saw more than a handful of Oswals. (In addition, some Srimalis from Saurashtra migrated not to Kenya, Tanzania or Uganda, but to other British protectorates such as Aden and Sudan). Thus while a tightly endogamous group – the Halari Oswals – settled in a single country and forming a highly-organized trading network, a more diffuse group of migrants – the Visa Srimalis from Jamnagar and a variety of other regions – dispersed themselves widely throughout the British colonies and protectorates and followed diverse occupational strategies.

Although secure and moderately well paid, employment in the Colonial administration obviously did not provide the Srimalis with opportunities for great economic advancement and capital reinvestment. Several of my Leicester Srimali informants expressed resentment that the less well educated Oswals had, through individual success and corporate solidarity, raised their status, power and prestige in East Africa (both as individuals and as a group), while the Srimalis with their better education and 'commitment' to East Africa had not. This difference in occupational strategy is probably the reason the Oswals settled only in Kenya, where they were able to develop a network of business contacts and information, while the Srimalis were free to seek government service in any of the three countries.

By the time independence movements gained ground in the East African countries, and the position of Asian traders and administrators began to look less and less secure, the Halari Oswals had taken on the appearance of a large, wealthy and successful group, comparable to the Patidars, while by comparison the Srimalis appeared small, dowdy and – crucially – lacking in investment capital. In the migrations that followed with the independence of Kenya, Uganda and Tanzania – moves that were largely forced upon the Oswals and Srimalis, as they were for all other South Asian groups (particularly in Uganda) – we again see a divergent pattern of migration. The richer Oswals migrated to London, the USA and Canada and, in some cases, back to Jamnagar – all destinations where East African-accumulated capital could be productively invested and where new business ventures begun. The Srimalis, by contrast, moved largely to Leicester, a town renowned among Gujaratis in East Africa as having a considerable Gujarati population and, more importantly, known as a town where unskilled jobs were freely available – an image presumably similar to that held in Pakistan of Bradford and other northern industrial towns. Many of the Srimalis were, of course, highly skilled administrators, bookkeepers, clerks and the like, but realized or quickly discovered that they were strongly disadvantaged in the British skilled labour market.

In Jamnagar, the returned Oswals established themselves within the city, or at least within the boundaries of what had by now become the wider municipality, land within the old city walls being unobtainable. There they established their own residential area, known as the Oswal Colony, which is today distinguished by large, ornate bungalows and two gleaming white Jain temples. Capital brought back (legally or illegally) from East Africa was used either to develop the family farm in the home village, through mechanical power and bore-well irrigation, or was speculated on the city's commodity futures market – a form of enterprise that has little or no need of office space, most of the business being conducted on the streets around the city's Grain Market. By the early 1980s when I arrived in Jamnagar, I found a degree of resentment towards the Halari Visa Oswals on the part of the Jamnagar Visa Oswals and the Jamnagar Srimalis who had remained behind. I was told that they were not 'real' or 'true' Jains, that they had no real aptitude for business and that the apparent economic and social success I could see all around me was but a fleeting stroke of corporate luck.

In short, I found a situation of flux, one where there has been jockeying for position among three major *jatis*, where the results of migratory strategies appear to have dealt some individuals – and thus their associated groups – good hands, while simultaneously dealing others bad hands. My evidence underlines the observations I made at the start of this paper, that the motivations for migration are by no means clear-cut, and the consequences by no means foreseeable. One thing is clear, however: anthropologists and others cannot hope to understand and describe migration if they persist in using models and language developed by economists, if they persist in seeing gross economic factors of 'push' and 'pull' as simultaneously descriptive and explanatory of complex processes which depend as much upon individual and corporate agency as they do upon wider and impersonal market forces.

REFERENCES

1. J. Abu-Lughod, 'Comments: the end of the age of innocence in migration theory', in B.M. du Toit and H.I. Safa (eds), *Migration and Urbanization: Models and Adaptive Strategies* (The Hague: Mouton, 1975), p. 201.
2. P.H. Gulliver, 'Myakyasa labour migration', *Human Problems in British Central Africa: The Rhodes Livingstone Journal*, 21 (1957), pp. 58–9.

3. For a rational choice model see B.M. du Toit, 'A decision-making model for the study of migration' in du Toit and Safa (eds), op. cit.
4. J. Eades, 'Introduction. Anthropologists and migrants: changing models and realities', in J. Eades (ed.), *Migrants, Workers and the Social Order* (London, Tavistock: 1987), p. 10.
5. The background ethnography to the following discussion is in M. Banks, *Organizing Jainism in India and England* (Oxford: Clarendon Press, 1992).
6. Abu-Lughod, 'Comments: the end of the age of innocence in migration theory', p. 205; P. Werbner, *The Migration Process: Capital, Gifts and Offerings among British Pakistanis* (Oxford: Berg, 1990).
7. E.R. Leach, *Pul Eliya: a village in Ceylon. A study of land tenure and kinship* (Cambridge: CUP, 1961), p. 300.
8. L. Dumont, *Homo Hierarchicus: The Caste System and its Implications* (complete revised English edn, Chicago: University of Chicago Press, 1980).
9. D. Quigley, *The Interpretation of Caste* (Oxford: Clarendon Press, 1993).
10. R.G. Fox, '*Varna* schemes and ideological integration in Indian society', *Comparative Studies in Society and History*, 11 (1969), 27–45.
11. But see L. Brydon, 'Who moves? Women and migration in West Africa in the 1980s', in Eades (ed.), *Migrants, Workers and the Social Order*.
12. See for example, V.A. Sangave, *Jaina Community: A Social Survey* (Bombay: Popular Prakashan, 1980), p. 86–7.
13. J. Tod, *Annals and Antiquities of Rajasthan*, vol. II, (London: OUP, 1920), pp. 603–5.
14. R.V. Russell, *The Tribes and Castes of the Central Provinces of India* (London: Macmillan, 1916), p. 161.
15. D.P. Khakhar, 'Castes and tribes in Kachh', *Indian Antiquary*, 5 (1876), p. 173.
16. C.A. Kincaid, *The Land of Ranji and Duleep* (Edinburgh: William Blackwood, 1931), p.16.
17. R.G. Fox, *From Zamindar to Ballot Box: Community Change in a North Indian Market Town* (Ithaca: Cornell University Press, 1969), p. 70.
18. Ibid, p. 77.
19. D.R. Mankad, *Jamnagarno itihas* [History of Jamnagar] (Aliabada, Jamnagar: Haribhai Research House, 1972), p. 106.
20 C. Fuller, 'British India or traditional India? An anthropological problem', *Ethnos*, 3–4 (1977), p. 96.
21. H. Spodek, 'Rulers, merchants and other groups in the city-states of Saurashtra, India, around 1800', *Comparative Studies in Society and History*, 16, 4 (1974), p. 450.
22. Ibid.
23. J. Zarwan, 'The social and economic network of an Indian family business in Kenya, 1920–1970', *Kroniek van Afrika*, 3 (1975), p. 220.
24. A.M. Shah, 'Division and hierarchy: an overview of caste in Gujarat', *Contributions to Indian Sociology (NS)*, 16,1 (1982), p. 21.
25. M. Michaelson, 'Caste, kinship and marriage: a study of two Gujarati trading castes in England' (PhD thesis, University of London, 1983), p. 93.
26. J. Zarwan, 'The social evolution of the Jains in Kenya', B.A. Ogot (ed.), *Hadith* [Special issue: *History and Social Change in East Africa]*, 6 (1974), p. 135.

27. Michaelson, op. cit., p. 92.
28. Banks, *Organizing Jainism in India and England*, p. 49, n. 3.
29 Banks, op. cit., chapter 4.
30 H. Spodek, 'Rulers, merchants and other groups in the city-states of Saurashtra, India, around 1800', *Comparative Studies in Society and History*, 16.4 (1974), pp. 448–70.
31. C. Cottam, 'City, town and village: the continuum reconsidered', in K. Ballhatchet and J. Harrison (eds), *The City in South Asia: Pre-Modern and Modern* (London: Curzon 1980), pp. 337–8.
32. J. Zarwan, 'The social and economic network of an Indian family business in Kenya, 1920-1970', p. 224.
33. N.K. Dumasia, *Jamnagar: A Sketch of its Ruler and its Administration* (Bombay: The Times Press, 1927)
34. J. Zarwan, 'The social and economic network of an Indian family business', p. 222.

7 The Movement of Indian Muslims to West Pakistan after 1947: Partition-Related Migration and its Consequences for the Pakistani Province of Sind

Sarah Ansari

This chapter sets out to examine the impact which migration from India, set in motion by partition in 1947, had on post-independence developments in the newly-created state of Pakistan. Partition-related migration was complicated from the outset by the sheer numbers involved and the political circumstances in which it was taking place. But it was made all the more complex because of the differences which distinguished and, to a large extent, increasingly separated the experiences of migrants moving from different parts of India to different parts of Pakistan, once the initial period of upheaval was over.

As a result, migration had a differential impact on the country as a whole. In some parts, notably the Punjab, the process of settling was fairly smooth and contributed towards the strengthening of the province in relation to the rest of the country. In other places, such as Sind and the city of Karachi, almost the reverse was the case, and the political unrest of more recent years is directly related to changes which the province underwent during the late 1940s and early 1950s. The period following independence thus offers a context in which to examine processes of integration and assimilation, and the factors which can encourage and inhibit these processes. By focusing on West Pakistan, with particular emphasis on the southern province of Sind, the chapter seeks to provide an appreciation of why and in what ways Indian Muslim migration to Pakistan has varied in its impact.

Large-scale migration in general and the movement of large numbers of involuntary migrants or refugees in particular have become a major feature of twentieth-century world history. The migration which accompanied the partitioning of British India in August 1947 certainly constitutes one of the largest and quickest mass-migrations seen this century. As in the case of similar population movements elsewhere, this transfer of population, quite apart from the immediate short-term dislocation involved, has contributed directly towards high levels of instability in the region. One of its more unavoidable long-term legacies has been the creation of circumstances in which ethnic tension has been the hallmark of the relationship between the original population of the region and the newcomers. Throughout the Indian subcontinent, migration has led to substantial levels of ethnic conflict, usually provoking resentment on the part of locals against the newcomers who are seen to be upsetting the existing balance of interests in the region or state.[1] Nowhere in Pakistan has this kind of tension been more pronounced than in the towns and cities of the province of Sind. Prior to 1971, the main source of ethnic tension within the country was that which resulted in the secession of East Pakistan and the creation of Bangladesh in its place. Since then, ethnic unrest has been most severe in Sind where Sindhis have come to form virtually a minority in their home province. Along with the realization that they are being increasingly outnumbered by other ethnic groups has come the belief that they are also losing out in terms of access to education, employment, land, political power and cultural identity. The result of the tension between Sindhis and migrants has been extreme eruptions of violence, deepening suspicion on either side, and the emergence of separate political parties representing both extremes. In many ways, the two groups remain as divided as they were at the time of partition and a migration which promised to bring an end to the problems of the Muslims of the Indian subcontinent has merely substituted old problems with new ones.

At least eight million Muslims crossed over into Pakistan from India in the years immediately following partition with equivalent numbers of Hindus and Sikhs moving in the opposite direction.[2] Migration on this scale took place as a popular response to the political decisions which accompanied the ending of British power and resulted in the creation of Pakistan, the first modern state to be set up on the grounds of religion. The Muslim League's Pakistan demand, issued as a counterweight to Congress's claims to authority at the political centre of an independent India, called for a 'homeland' for the Muslims of the subcontinent on the grounds that they constituted a distinct nation and had the right to a separate state at independence. Pakistan was, therefore, intended to protect Muslims against the perceived dangers of living in an independent Hindu-dominated India. While a historical question mark hangs over whether or not the League leader, Muhammad Ali Jinnah, intended Pakistan to become a separate state in the way that it eventually emerged, there is no doubt that the extent of the territory which became Pakistan fell short of his and the League's final expectations. Deadlock in negotiations with Congress together with growing communal tension resulted in a British plan to partition India, including the provinces of the Punjab and Bengal, which, as they contained slight Muslim majorities, the League had expected to receive in full in any territorial division of the spoils. Jinnah, therefore, was far from happy with the 'truncated' and 'moth-eaten' state which he was offered, but, with the effective alternative of conceding power completely to the Congress, the League finally accepted this option in the summer of 1947.[3]

What turned out to be completely unexpected, however, was the size of the migration caused by partition. According to the 1941 census, Muslims had then comprised only 24.3 per cent of the total population of the subcontinent. It was the concentration of Muslims in certain parts of India which made partition possible six years later. Out of a total of 435 districts or other comparable divisions in India in 1941, there were 76 whose population was more than half Muslim. Had these districts been scattered throughout India, partition would have been a non-starter. But, instead, they were grouped into two clusters, one in the north-west and the other in the north-east of the country. Not a single Muslim-majority district existed apart from these two clusters, nor did either cluster completely contain a Hindu-majority district. Consequently, the two clusters, embracing some 56 million Muslims in 1941, 'formed the geographic and demographic reality that made the idea of Pakistan feasible'.[4]

While the geographic concentration of Muslims was great enough to give weight eventually to the political arguments in favour of Pakistan's creation, it was not however great enough to ensure that either the new country would contain all the Muslims in India or that it would contain no non-Muslims. The existence of 38 million Muslims outside the concentrated areas and 20 million non-Muslims within them meant that instead of one minority problem – that of a Muslim minority living in a Hindu-majority state – there would be two, with considerable numbers of Sikhs and Hindus living in a Muslim state. One way around this problem might have been a mutual exchange of populations, but such a plan would have been virtually impossible to organize, notably because of the difficulties of matching the people exchanged so as not to upset the economy on either side of the border as well as actually inducing people to move in the first place.

Such obstacles to a 'planned' migration, however, were overwhelmed by the spontaneous massive cross-migration that actually took place. Despite warning signs, such as the Hindu attacks on Muslims in Bihar in October 1946 which had led most politicians to fear that there would be some loss of life as a result of partition, few observers envisaged that the creation of Pakistan would trigger off the extremely bitter communal reactions which provided much of the impetus for migration.[5] Thus, it turned out to be almost as much as a surprise to British officials and the respective governments of India and Pakistan as it was to the bewildered refugees themselves. The suddenness of the political deal which ended British rule in India made all attempts at a planned exchange of population an impossible task. Within six months of partition, hundreds of thousands of Hindus, Sikhs and Muslims had died, caught on the wrong side of the new borders. Many were killed by a combination of so-called 'natural causes', such as disease, hunger and exhaustion, but many others were victims of intense communal violence. Millions of rupees-worth of property was also destroyed and the social and economic organization of the affected areas was seriously dislocated. Both new governments were faced with large refugee populations who required immediate assistance and lengthy rehabilitation programmes at a time when both states were in the process of trying to establish themselves with limited resources at their disposal. The cross-migration inevitably became a matter of official concern to both governments because it created almost insuperable difficulties in connection with evacuation and transportation, the provision of shelter, food, means of livelihood, public hygiene and medical relief. In addition, the Pakistan authorities faced the problem of a severe shortage of experienced administrative staff. The migration to India en bloc of Hindus and Sikhs who had previously filled a majority of senior posts in the local administration of what became West Pakistan nearly paralysed its administrative machinery.[6]

Problems of this kind contributed to Pakistan's failure to secure accurate statistics of the refugee movement during the early chaotic months of the country's existence. It is not surprising to find that the estimates of the refugee movements vary a great deal depending on the agency which attempted to make them. The discrepancies between the Pakistani and Indian estimates were particularly great – each country attempting to claim that it had received a considerable net surplus of refugees over evacuees and each country insisting that their respective nationals had been forced to abandon the greatest amount of wealth during the evacuation. The overwhelming majority of migrants who crossed into West Pakistan, however, were poor, predominantly Punjabi, cultivators fleeing communal violence further east. By the end of 1947 nearly all Muslims from East Punjab had migrated westwards. There were also substantial numbers of migrants from Delhi and surrounding districts, as well as Muslims from the United Provinces and Bihar. Muslims living further south, in the princely states of central India such as Hyderabad, or in western India such as Bombay and Rajasthan, also made their way in large numbers to West Pakistan. In contrast, the migration pattern in Bengal during the first two years following partition was largely a one-way movement – Hindus leaving East Bengal for the new Indian state of West Bengal – and it was not until a resurgence of communal activity in 1950 that large numbers of Muslims crossed into East Pakistan from India. The main reception centres in the immediate aftermath of partition, therefore, were in West Pakistan and in the Punjab in particular. Lahore had the largest refugee camps, while 1.5 million refugees alone occupied the four West Punjab districts of Lyallpur, Montgomery, Shahpur and Multan.[7] The southern province of Sind also experienced a huge influx of migrants which placed equally large strains on local resources. Karachi expanded from about 400 000 in 1947 to over 1.3 million in 1953, placing a huge strain on the life of the city where building activity had almost totally ceased during the Second World War. (See Table 7.1)

In overall terms, the Muslim refugees who migrated to West Pakistan represented only about 20 per cent of the total population of the new country, and yet the long-term significance of their arrival was great. Economically, socially, politically and out of all proportion to their numbers, they proceeded to contribute to the shaping of subsequent developments in the parts of Pakistan where they settled. Nowhere was this more obvious than in the city of Karachi and its hinterland, the southern Pakistani province of Sind. Although not the main recipient of migrants from India in terms of total numbers, nor the scene of much communal violence prior to independence, the influence of partition proved to be far-reaching. Before partition, Sind had been a relatively sleepy backwater of British India. Its population of under five million was overwhelmingly rural: apart from Karachi on the coast and a handful of minor cities further inland, Sind's urban profile consisted of small trading centres and market towns. There was very little industry and Karachi itself was first and foremost important as an entrepôt centre for goods being moved to and from parts of India further north and east.[8]

Table 7.1 Approximate number of refugees in each province of West Pakistan to 1950

Province	
Punjab	5 567 000
North West Frontier	47 700
Sind	889 000
Baluchistan	15 000
Karachi	956 000
Bahawalpur*	351 500
Khairpur⁺	23 900

* subsequently incorporated into the Punjab

⁺ subsequently incorporated into Sind

Source: Economy of Pakistan', Ministry of Economic Affairs, Government of Pakistan, n.d., p. 386.

To some extent, Sind had begun to change before independence. This was largely as a result of increased irrigation which had speeded up commercialization of agriculture and the rate of change in the countryside during the interwar period. But the speed of this change was considerably increased after 1947. Agriculture became much more commercial than before. Sind also acquired a much firmer industrial base, particularly in Karachi which became Pakistan's leading industrial centre. In both processes, migrants played a central part in introducing innovation and taking the entrepreneurial initiative. As a result of these developments, there was an important restructuring of the economic and demographic balance of the province with accelerated growth in the relative importance of urban centres.

The migrants who settled in Sind were also, on balance, a more educated group than the province's original Muslim population which had lagged far behind the province's former Hindu population in terms of educational attainment and the jobs which accompanied it. The position of Muslim women also underwent subtle but significant change. The overall trend during the decade following independence was a steady movement towards greater participation by women in the public life of the province led by migrant women from more urban backgrounds. Female educational facilities expanded, as did the participation of women in the workplace, albeit often within traditionally acceptable spheres of employment such as education and medicine. Most symbolic of this newly acquired 'freedom' was the frequent discarding of the *burqa* or veil upon arriving in Pakistan and the emphasis placed on the role which migrant women could play in helping to create a new future for themselves and their families.[9]

At the same time, many migrants brought with them a style of Islam which emphasized the scriptures and thus contrasted with Sind's strongly Sufistic and more syncretic religious tradition. In consequence, the years following partition saw the growing popularity among refugees of religio-political parties, when disillusionment with what both local society and the machinery of the state had to offer encouraged them to look for religious solutions to their problems. On a lighter note, other new habits began to take root in the province. The twin vices of the chewing of betel-nut leaf or *pan* and tea-drinking accompanied the migrants to Sind. By the early 1950s, the authorities had become so conscious of the burden of importing *pan* leaves that attempts were made to cultivate the crop locally – but without much success. Cricket and hockey began to replace football as the dominant local sports, reflecting the interests of the migrants from India. The popularity of locally-produced Urdu films, not to mention that of films imported from India, formed part of the wider process by which Urdu, the language of north Indian Muslims, established itself in its new cultural home. While English remained the language of the bureaucracy, Urdu rapidly became the lingua franca of migrant-dominated cities such as Karachi, and so its dominance over provincial languages such as Sindhi was assured even early in the 1950s.

The kinds of changes which were set in motion in Sind after August 1947 were not restricted to that province alone. From the point of view of agriculture, industry, urbanization, education, social change, even religious practice, developments in the Punjab followed a similar pattern.[10] Yet the outcome for each province was significantly different in terms of the kind of society which was produced. Unlike Sind, the Punjab has exhibited few of the ethno-political problems from which Sind has suffered: it has become a much more integrated society compared with the ethnic 'chaos' which has characterized Sind. This difference, in large part, can be traced to the fact that migration had a differential impact on the provinces making up Pakistan. The capacity of different provinces to deal with and absorb their quota of refugees played an important part in determining how smoothly the process took place. It was the Punjab, not Sind, which bore the main brunt of the initial refugee avalanche. The point has already been made that problems exist in calculating the exact number of refugees, largely because no records of the mass movements were compiled on the Pakistani side during the first two years following partition. But there can be no doubt that the number of migrants to the Punjab far exceeded the total number for the rest of the country. However, the Punjab was the largest province in West Pakistan and it was also the strongest economically. Sind in contrast was smaller and had fewer resources. The eventual arrival there of over a million refugees represented about one in four of the population of the province compared with the national average of one in six. The Punjab and Sind also differed in the pattern of refugee arrival which they experienced. In Sind the cycle of refugee arrival repeated itself for far longer than was the case elsewhere and effectively dragged out much more the practical problems associated with the rehabilitation process. In 1953, migrants were still arriving in formidable numbers in Sind while, in marked contrast, the trickle crossing from India into the Punjab had all but dried up by this time.

Another equally important distinction was the variety of migrants who sought to settle in Sind. Most of the refugees who ended up in the Punjab were Punjabis themselves, nearly two-thirds at least started off as individual cultivators who could continue as such once the process of rehabilitation had got under way. The economic and cultural dislocation involved in their migration was reduced by the fact that they were still operating within an environment and an economic framework to which they were accustomed. This helped to ease their acceptance by West Punjabis despite some local resentment at what seemed to be the preferential allocation of resources to refugees. The migrants who ended up in Sind were far more heterogeneous. They also had much less in common with the host community than their counterparts in the Punjab. This was the case whether the migrants were 'overspill' East Punjabis, forced to opt for Sind when evacuee land was no longer available in the Punjab itself, urbanized Urdu-speakers from north India, or Muslims from western India with strong commercial connections. Some 70 per cent of the second of these categories settled in Sind's towns and cities. To some extent, this choice of province had been made for them when, during the initial panic to leave India, many Muslims from the United Provinces, for instance, took trains which were routed through Jodhpur to Sind in order to restrict the number of refugees entering the Punjab and avoid the communal violence which characterized Punjab during this period. Once the shape of the new state had begun to clarify itself, the trend was intensified by the fact that, as a predominantly urban-based group heavily involved in government service, professional occupations and skilled crafts, this particular group of migrants saw the new federal capital of Karachi as the place most likely to meet their employment needs. Economically and culturally, these two groups were distinct both in relation to each other and in relation to local Sindhis. Apart from anything else, they spoke different languages. So, in contrast to the Punjab, very basic obstacles to communication threatened from the outset to slow down the rate of integration in Sind.

Divisions between the '*muhajirs*', as the largely Urdu-speaking refugees came to be known, and the so-called 'ethnic' Sindhis were reinforced by the fact that the former were predominantly urban while the latter remained overwhelmingly rural. Although large numbers of formerly rural Punjabis became workers in the expanding towns and cities of the Punjab, the gap between the two worlds was less pronounced than in Sind. For a start, the process of rehabilitation proceeded much more quickly in the Punjab. By 1954, a scheme to grant provisional permanent rights over allotments of evacuee land to migrants had been virtually completed in the Punjab whereas in Sind progress lagged far behind.[11] Under these relatively more settled conditions, agricultural production in the Punjab became more commercialized than in Sind, and in this way, the links between town and countryside were strengthened by economic forces which, while present in Sind, did not exist there to nearly the same extent. Even such basic factors as the relative ease of communications between countryside and town in the Punjab were reinforced by the superior network of roads and railways which linked villages into a wider framework of connections. Much of the western Punjab had been systematically developed with the construction of 'canal colonies' by the British when irrigation in the region was greatly expanded at the turn of the twentieth century. The outlook of the average Punjabi cultivator was also affected by the enduring ties which linked the Punjab with the military. A tradition well-established during the British period of recruiting from the supposedly hardy peasant stock of the province and then resettling ex-soldiers on the land continued into the post-independence period. This meant that many rural families in the Punjab had direct experience of a wider world than was the case for Sindhis who carried on being regarded, as under the British, as unsuited to military service.[12]

The first decade following partition represented a crucial period in the forming and hardening of the perceptions of the different communities living in Sind. On the one hand, many Sindhis began to resent the way in which other groups seemed to be benefiting at their expense. On the other hand, instead of growing closer to the host community, many refugees came to see their interests as being distinct and even special. The basis of their relationship was decidedly shaky, and it grew more unstable during the years after 1947. The initial Sindhi response to the plight of the arriving migrants was welcoming. There was a sense of shared responsibility for what was happening, in that Sindhi Muslims prided themselves on having played a crucial role in advancing the Pakistan cause, and with this came sympathy on a humanitarian level for the migrants and their accounts of experiences en route to Pakistan. But as the numbers of refugees climbed and their economic impact was felt more keenly, the welcome wore thin and was often replaced by resentment.[13] Sind, like its neighbouring provinces, experienced the practical difficulties of looking after so many destitute people, and under these circumstances competition for resources such as land and the scramble for jobs in urban areas combined to produce a gradual coolness towards the migrants on the part of certain sections of the Sindhi population.[14] Even before independence there had been some dissatisfaction at the presence of non-Sindhis in the province, encouraged to migrate there in increasing numbers by British policies connected in particular with the expansion of irrigation and agricultural production in Sind. The authorities, to recoup the cost of constructing new irrigation systems, had actively assisted the arrival of Punjabi settlers, whom they regarded as more hard-working and generally more competent compared with Sindhis who tended to be seen as lazy and afraid of any kind of strenuous activity.[15]

This situation was not helped further by the fact that migrants often expected a degree of preferential treatment, regarding it as just compensation for what they had been forced to leave behind in India. The name by which they were increasingly known – *muhajir* – widened this gap in perceptions. Whereas a refugee is pushed largely by circumstances, a *muhajir*, it is said, makes a positive decision to migrate, and hence might well consider it reasonable to receive respect, including material benefit, from the wider Muslim community. In reality, the majority of *muhajirs* were acute migrants – refugees who had seen little option but to leave for Pakistan: however, once in their adopted homeland, it was possible for different interpretations of their flight to come into play. *Muhajirs* often could not understand that Sindhis were facing the difficult problem of how to reconcile provincial identity with the reality of an increasingly centralized Pakistani state dominated by non-Sindhi interests. Hence, *muhajirs* perhaps expected more of Sindhis than was realistic under the circumstances. It was not long before newspapers recorded outbreaks of friction between refugees and local people, and mention of them only disappeared once editors began to co-operate with a government request not to publicize such incidents. Politicians exhorted both sides to coexist peacefully: for Sindhis to think of themselves as 'Ansars', the original inhabitants of Medina who had welcomed the Prophet Muhammad and his companions, the original *muhajirs* after they had left Mecca; and for *muhajirs* to consider the Ansars as their elder brothers and, if there were any grievances against them, to forgive them as was done in the case of elders. Both sides, in their own interests as well as those of the state were asked to think of the migrants as Sindhis for talk of Sindhi and non-Sindhi was regarded as suicidally divisive.[16]

Official Government of Sind reactions to the arrival of the migrants, however, played their own part in helping to establish suspicions among *muhajirs* that perhaps they were not so welcome after all. For instance, when by September 1948, it was realized that there was a surplus population of refugees in the Punjab who could not be allowed to stay without severely jeopardizing its economic life, the central government took the decision to move nearly 200 000 of them to Sind.[17]

Much discussion was generated and a great deal of pressure had to be placed on the Sind authorities before they grudgingly accepted this extra quota. While their objections were valid to a certain extent, in that most of the non-Muslims who had migrated from Sind were non-agriculturalists and therefore it would be difficult for the province to take more of the cultivators who made up the bulk of the Punjabi shipment, their reluctance created a negative impression in *muhajir* minds. This was subsequently reinforced by the Government of Sind's response to an upsurge in migrant numbers which started during the first half of 1950 as a reaction to heightened communal tensions in India. The authorities objected to the influx, which reached the rate of 3000–4000 a day at its peak, on the grounds that it placed enormous strains on the province's ability to handle their arrival when the resettlement of earlier refugees had hardly got under way. Initially, they simply called on refugees to return to their homes in India. When this did not work, they withdrew facilities, in the shape of camel-trains and drinking water, for refugee reception at the border. Crossing into Pakistan at that time of the year when temperatures could reach 120°F and which involved travelling across some 40 miles of barren sandy desert without shade, shelter or water, proved daunting. Not surprisingly, by early June 1950, this measure, together with the imposition of a permit system, dramatically reduced the number of migrants. The combined effect of these official moves proved deeply worrying to *muhajirs* who felt that they had good reason to fear that they might eventually be forced to go back to India against their will.[18]

All these developments in Sind helped to reinforce the psychological as well as the physical distance separating countryside and town and Sindhi and *muhajir*. Most *muhajirs* had a set of priorities which had little in common with most ordinary Sindhis. Well into the 1950s, hundreds of thousands of refugees had not been resettled. Karachi was especially bad: some 250 000 people were still living in 'improvised dwellings' in 1953, and over 100 000 people were in similar conditions in 1958. The worst of the so-called colonies consisted of haphazard groupings of small huts constructed from whatever material was available such as bamboo, thatch and hessian rags. Water and sewerage arrangements were usually non-existent. Small clusters of huts were also found along the walls of permanent buildings and main thoroughfares. The monsoon rains annually caused havoc, washing away homes and spoiling possessions and food supplies.[19] The migrants' primary aim therefore was to obtain adequate housing; and government's inability or incapacity to deal effectively with the problem generated harsh criticism. Refugee organizations sprang up to protect *muhajir* interests and many were very critical of the administration. One major grievance was the 'usurping' of evacuee property by influential locals. A popular focus of migrant activity was the co-operative housing society which sought to purchase land for residential development. Many societies found their plans frustrated by bureaucratic delay, which meant that much of the 1950s was spent bickering over housing schemes. Government schemes were even slower to produce results. The economic hardships of the first decade following partition increased economic insecurity, unemployment rose, and standards of living fell. The 1950s were a period of bitterness for many *muhajirs* who watched the rich become richer as inflationary spirals forced down their own real incomes.

For the majority of Sindhis, however, the problems of urban life were fairly remote. The main issue for them was land: either to hang on to it as in the case of landlords, or to get hold of some if they were tenant cultivators or landless agricultural workers. The 1950s as a result saw a long-running saga of calls for and half-hearted attempts at land reform; but the strength of the landed interest in provincial politics meant that, when land reform did come, it was limited to only certain kinds of estates and did not affect the huge holdings for which Sind was notorious.[20] Redistribution of land was not as significant an issue for refugees who, if they were Urdu-speaking, concentrated on their own urban concerns, while Punjabis who had settled in the countryside often saw themselves as targets of local resentment since they had been allotted land which might otherwise have been given to Sindhis. This source of tension was heightened by the allocation of land to armed forces personnel who were predominantly from the Punjab and which helped to reinforce long-held Sindhi suspicions about all things Punjabi. Meanwhile, many Sindhi landowners chose to interpret land reform programmes as schemes devised by refugee interests to destroy the basis of established power in the province.

Polarization became equally apparent on the economic front. Many of the migrants, especially those who settled in the smaller towns of the interior of Sind, came to fill the vacuum left by departing Hindus and in the process inherited some of the resentment which had been directed towards them by Sindhi Muslims. This was reinforced by the way in which migrant families came to dominate industrial and commercial life in the province. By the end of the 1950s, Pakistan's leading industrial families included a substantial number who had migrated to Sind in or soon after 1947. At the other end of the economic scale, many poorer migrant families found themselves working in factories providing the labour force for this industrial growth. An atmosphere of competition emerged in the service sector which came to be seen as the domain of the Urdu-speaking migrant. Government posts and other high-status occupations seemed to be filled with *muhajirs* who tended to be better qualified and consequently better placed to compete for these jobs. From some Sindhi perspectives, the migrants were becoming a new group of adversaries, replacing the old Hindu élite in the province and restricting opportunities for them once again. There was even some talk that Hindu 'domination' had been preferable to this new situation for Sindhis. Hindus had at least spoken the same language and shared a common history – which the migrants did not.[21]

A shared religion, the reason for Pakistan's creation, did not automatically bring the two communities closer together. Although all were

Muslims, differences between their styles of Islam could produce friction. Many *muhajirs* simply looked down on Sindhis for what the former regarded as unhealthy syncretic tendencies in Sindhi Muslim religious practices. As Islam was so closely identified with the founding of Pakistan, religion came to be used by the new state as a means of cementing over divisions within the country; but when it was a form of Islam which reflected the traditions of northern India its effect could be divisive in somewhere like Sind. Islamic unity, for example, was used as an argument to persuade Sindhis to accept Karachi's becoming a federal territory: thus religion was placed in direct conflict with provincial loyalties. The importance of their kind of Islam, at least symbolically, to many migrants was illustrated by the predominance of religious men among their representatives. It continued in the shape of the backing which these leaders gave to campaigns to ensure that the constitution respected the fact that Pakistan was supposed to be a Muslim state. The inadequacies of the state, combined with the growing belief that their interests would be best served by pursuing a political course which emphasized Islam as a means of applying pressure for change, resulted in high levels of migrant support for religio-political organizations. By the end of the 1950s, for example, the Jamaat-i Islami had built up the main basis of its support in the largely Urdu-speaking, migrant-dominated cities of Karachi and Hyderabad. For Sindhis, however, the Jamaat and its counterparts remained organizations dominated by and reflecting largely outside interests.

The perception in Sindhi minds that they were under threat of being dominated by 'outsiders' was reinforced by cultural developments which seemed to jeopardise Sind's identity. Most symbolic was the fate of the Sindhi language which became an important focus of discontent among the more politicized Sindhis. Sindhis increasingly found that they needed to learn Urdu but the reverse was not true for Urdu-speakers; not being an Urdu-speaker seemed to put Sindhis at a disadvantage which they could not easily overcome.

More generally, the two communities had political agendas which barely seemed to overlap. Sindhi politics were the preserve of large landed élites. Even urban-based Sindhi politicians normally had absentee landlord interests. In the past, the indigenous power structure of the Sindhi countryside had tended to coalesce to prevent outsiders from disturbing the status quo; but the scale of the migrant influx was sufficiently large to threaten this state of affairs and so Sindhi politicians felt a pressing need to protect their position. Migrants, for their part, were frustrated for much of the 1950s by a lack of adequate representation even bearing in mind the shortcomings of Pakistan's political system at the time. Their feelings of being distinct led them to consider themselves to be in a separate political category from other people in the province. Throughout Pakistan, there were demands for separate seats for refugees in provincial assemblies; but whereas in the Punjab refugees were automatically allocated seats formerly filled by non-Muslims, refugees in Sind did not obtain them. The inclusion of a lone refugee minister in the provincial ministry was interpreted by most *muhajir* observers as at best a half-hearted attempt to reflect their interests or to meet their special needs. The main focus for their political activity not surprisingly became the Karachi and Hyderabad municipal corporations and parallel city political organizations. But divisions within these bodies between migrant and non-migrant persisted and were reflected in the kinds of factional divisions which emerged in these cities.[22] The concentration of migrants in Karachi, the seat of the federal government, reinforced the tendency of *muhajirs* to be identified with the central authorities. Migrants tended to support moves by the centre to extend its control over the city which led it to being separated from Sind in July 1948. Sindhi reactions to this move were vehement and Karachi's separation acted as a potent symbol of the antagonism felt in Sind towards the central government. *Muhajir* support for the scheme was seen to be 'proof' that they did not have Sind's 'best interests' at heart. Indeed many *muhajirs* approved of the transfer precisely because they wished to operate within a framework in which there was more possibility of their having a prominent political voice. *Muhajir* identification with the centre, despite all its shortcomings, at a time when centre–province relations were poor did not endear them in Sindhi eyes when the centre was often interpreted as the extension of Punjabi interests.

These developments suggest that during this crucial period in the forming and hardening of perceptions about each other, the main communities, which now made up Sind's population, grew further apart rather than coming closer together. For a combination of reasons, they failed to identify mutual interests and construct a strong enough common identity vital to the creation of a more integrated society. This failure was in striking contrast to the situation in the Punjab during the same period. As the Punjab's influence over the country as a whole increased, so 'Punjabi' and 'Pakistani' identities tended to merge to the point that most Punjabis saw no discrepancy between the two. In contrast, the Sindhi identity seemed to be totally submerged by the Pakistani one, something which the relative weakness of the province could not resist. The introduction of One Unit in 1955 when the provinces of West Pakistan were merged to create parity with East Pakistan only seemed to confirm Sindhi fears that the centre was dominated first and foremost by non-Sindhi/Punjabi interests. Many Sindhis had rejected the idea of One Unit on the grounds that it was against Sindhi interests from the time that it was first broached in the late 1940s.[23] Migrants' support for the move was interpreted as yet more proof of their design on power locally as well as a share in national power.

In the same way the imposition of martial law in 1958, which effectively ended the first decade of independence, highlighted the difference in outlook of politicized Sindhis and migrants. While the former viewed it as the extension of alien, albeit largely Punjabi, military and civil bureaucratic control over their province, many *muhajirs* greeted it with a large measure of relief, such was their disillusionment with the civilian authorities' ability to tackle their continuing problems. As so many middle-class *muhajirs* formed an important element within the bureaucracy, they appeared to be among the direct beneficiaries of the loss of political rights in Sind and the country as a whole. In this area as in others, *muhajir* interests overlapped with those of the centre while Sindhi ones did not, reinforcing still further the hardening political divide between migrants and Sindhis themselves.

By the end of the 1950s Sind, although considerably affected by the impact of partition-related migration to the province, had yet to come to terms with the problems, both real and potential, which were partition's legacy. In contrast, once the military were in overt power from 1958 and the influence of Punjabis was increasing in relation to *muhajir* involvement in the country's bureaucracy, the Punjab was poised to take advantage of the emphasis on economic development and modernization which became the watchwords of President Ayub Khan in the 1960s. Sind to some extent benefited from the economic imbalance between East and West Pakistan which increased during this period but the 'gains' which it made were largely limited to the further industrializing of Karachi and had much less impact on the province as a whole. Similarly, the completion of additional irrigation facilities provided a boost to agricultural production but it remained debatable whether or not it was Sindhis or 'outsiders' who gained most from the changes. There was always the perception on the part of Sindhis that non-Sindhis were the main beneficiaries of any development which did take place.

Concentrating on 'migrants versus Sindhis' does of course somewhat simplify reality for the sake of clarity: class interests could and did divide ethnic communities and equally bring members of different ethnic groups together on particular issues. Likewise, divisions within the migrant community, for instance, between Punjabis and non-Punjabis, were always significant, to the extent that by the late 1950s, the term *muhajir* had become primarily identified with Urdu-speaking migrants. Nor has ethnic tension been confined to the Sindi-*muhajir* split. For example, the growing presence of large numbers of Pathans from the country's North-West Frontier Province, stimulated by the city's expansion, contributed directly to the serious outbreaks of ethnic violence which erupted in Karachi in the late 1950s.[24] This particular ethnic divide resurfaced in bloody fashion again in the 1980s.[25]

However, the Sindhi-*muhajir* problem does differ from the other kinds of ethnic confrontation present in Sind for, unlike people from other parts of the country such as Pathans and Punjabis (even those originally from the Indian side of the border) who can be called upon in theory to return to their own provinces, *muhajirs* cannot really be expected to return to India. Theories on the integration of involuntary migrants such as refugees have tended to stress the concept of assimilation as the yardstick by which to measure the degrees of difference between so-called 'host communities' and the migrants. Cultural assimilation was seen as the first stage when the universal values of the receiving community were incorporated into the migrant's lifestyle in order to minimize differences with their adopted society. Social assimilation was supposed to follow when the migrant moved into a face-to-face interaction with wider society on an equal basis. Assimilation was said to have occurred when this two-stage adjustment had been successful and the migrant was accepted by and felt 'belongingness' to the receiving society. This model, however, proved to be Western-centric even in the society in which it was formulated. Social scientists working in the Israeli context of the late 1940s and early 1950s produced theories which later had to be modified in the light of the realities connected with mass immigration. Later studies instead concentrated on the possibility and probability of cultural ethnicity being retained by migrants, resulting in a slowing down of the rate and kind of assimilation envisaged in the original theoretical models.[26]

In Sind, these two stages of cultural followed by social assimilation seem to have been largely missing from the local equation. The circumstances in which Pakistan was created meant that for many migrants there was never any question of their adopting the values of their host society – in this case Sindhi society – because of their own relationship with what the concept of Pakistan was supposed to mean. They arrived there with their own ideas and generally saw themselves to be the embodiment of the new state. If anything, they expected local society to adjust to them rather than adjustment being the other way round. The major vehicle for cultural assimilation in Pakistan has been religion rather than any kind of regional identity and it has been the Islam of the migrant rather than local religious practice which has been dominant. Similarly, while social assimilation could be considered important in Western societies, its significance is bound to be much less in places like Sind where, as elsewhere in South Asia, small-scale social groupings are still accepted and powerful features of the social landscape. Indeed, a particular group can be totally assimilated in political and economic terms but by choice have minimal interaction in the social sphere. So while studies of migrant assimilation in the West often measure the degree to which migrants retain contact with their cultural group and assume that assimilation has not properly taken place where strong ties are found, most groups in non-Western societies such as South Asia are almost expected to maintain strong links with their own communities.

These shortcomings of existing theories would suggest that other ways of considering the process of integration following migration, or lack of it, are required to understand the situation which developed in somewhere like post-partition Sind. In particular, it is necessary to take into account the kinds of tensions generated by the failure of the new, in this case Pakistani, state to create for itself the shared political identity so necessary for dealing with the stresses of 'nation-building' and to satisfy the expectations of all its people. Under circumstances such as these, differences which might otherwise have been more or less tolerated became potent sources of tension: and competition and resentment became the enduring hallmarks of the relationship between ethnic groups in parts of Pakistan such as Sind which lacked meaningful political power. As a result, sub-state nationalities have become so important that many *muhajirs* – originally the group most obviously connected with the Pakistani ideal as a result of their migration to the country – chose during the 1980s to communicate their perceived needs in the language which had become the dominant means of political expression in Pakistan, that of ethnic-based politics. Hence, the wide-scale support which developed during this decade for the *Muhajir Qaumi Mahaz* (MQM) which talked of migrants forming a fifth 'nationality' within Pakistan to add to the Punjabi, Sindhi, Baluchi and Pathan.

The emphasis involved in the way in which historians have tended to look at Muslim separatism, the demand for Pakistan, and the creating of the new state, has helped to produce an impression of a fairly uniform political response on the part of the Muslims of the Indian subcontinent to the prospect of independence Concentrating in this way above all on the coming together of Muslims which produced its climax in the shape of the separate state of Pakistan symbolized by the migration which accompanied partition, has effectively obscured the reality of the differences which separated Muslims themselves at the time. Partition-related migration brought these Muslims together. Where differences were slight, the process of settling proved successful. But where communities had relatively little in common, economic, political and social circumstances combined to keep them apart. Thus, while the changes which took place in Sind after partition altered Sind in ways which cannot be undone, their long-term outcome has been the creation of a province in which both migrants and Sindhis have felt dissatisfied with the new arrangements. Their practical effect has been to hinder processes of integration while making demands for assimilation on both sides all the more vocal, with, in effect, Sindhis calling on migrant families to become 'more Sindhi', and migrants insisting that Sindhis replace their provincial identity with a 'more Pakistani' one.

NOTES AND REFERENCES

1. Myron Weiner, *Sons of the Soil: Migration and Ethnic Conflict in India* (Princeton, 1978), examines the clash between migrants' claims to equal access and the claims of local groups to protection by the state within the Indian context.
2. For example, Joseph B. Schechtman, *Population Transfers in Asia* (New York, 1949), pp. 18–19, estimated the number of people entering Pakistan at about six million with some five to six million leaving; and later arrivals eventually pushed the total up to at least eight million, if not considerably more.
3. See Ayesha Jalal, *The Sole Spokesman: Jinnah, the Muslim League and the Demand for Pakistan* (Cambridge, 1984) for coverage of the League's campaign for a separate Muslim state.
4. Kingsley Davis, *The Population of India and Pakistan* (Princeton, 1951), p. 196.
5. Dow to Wavell, 22 November 1946, L/P&J/5/181, IOLR (India Office Library and Records).
6. The Refugee Problem in Pakistan', US Embassy, Karachi to State Department, Washington, Despatch No. 976, 28 April 1953, 890D.411/4-2853, USNA.
7. See 'Summary of the migrations of 1947–8', n.d., Mudie Papers, MSS/Eur/F/164/47, IOLR.
8. Railway construction from the end of the nineteenth century had linked Sind with Quetta, Lahore and Bombay, increasing the volume of trade through Sind tremendously and helping Karachi to develop into one of the subcontinent's major ports, see H.T. Sorley, *The Gazetteer of West Pakistan: The Former Province of Sind* (Karachi, 1968), p. 588.
9. For example, the role which was urged on women by a newspaper article by Begum Jalil Asghar entitled 'The Women's Movement in West Pakistan', *Pakistan Times*, 27 March 1958.

10. For a study of the impact of refugees on the industrial expansion of a Punjabi town, see Muhammad Waseem, 'Urban Growth and Political Change at the Local Level: the Case of Faisalabad City, 1947–75', in Akbar S. Ahmed, ed., *Pakistan, the Social Sciences' Perspective*, (Karachi: Oxford University Press, 1990), pp. 207–28.
11. See Foreign Service Despatch No. 67, 3 August 1954, 890D.411/8-354, USNA.
12. Sind had acquired an 'unwarlike' reputation in British military eyes which dated back to at least the First World War, if not earlier; see Political Department File No. 350 (1919), vol. III, comp. 20, n.p., CSR.
13. Resentment was increased by incidents which highlighted the growing hostility between refugee and non-refugee groups, such as the riots which took place when refugees attempted to occupy evacuee land, see *Dawn*, 31 March 1948, p. 1.
14. As early as January 1948, for instance, the Sind Muslim Students Federation called for 'Sind for Sindhis'. A refugee student organization promptly responded by arguing that while the people of Sind should enjoy a prominent position in the life of the province, they should not forget that the country had been established primarily because of the efforts and suffering of people forced to leave their homes and take asylum in Pakistan, see *Dawn* (Karachi), 27 January 1948, p. 6.
15. Lawrence to Birkenhead, 6 May 1926, MSS/Eur/D/70s/15, p. 26, *IOLR*.
16. An agency set up by the central government to help in the rehabilitation work was the Sind Field Publicity Unit. The members of this unit moved throughout Sind to maintain discipline among the refugees and sustain their morale. They were a link between the refugees and the local population, and aimed at creating 'brotherly feelings' between the two groups. A free fortnightly journal, *Nai Zindagi* ('New Life') was also published by the government for the benefit of refugees and local Sindhis in two separate Urdu and Sindhi editions. Its object was to educate refugees in the new way of life they were to lead in this province and to help local people in understanding the ways of their new neighbours, see US Embassy Karachi to State Department, Washington, Desptach No. 104, 31 July 1950, 890D. 411/7-3150, USNA.
17. US Embassy Karachi to State Department, Washington, Despatch No. 104, 31 July 1950, 890D.411/7-3150, USNA.
18. US Embassy Karachi to State Department, Washington, 'Notes on the Refugee Situation – West Pakistan', 12 June 1950, 890D.411/6-12250, USNA.
19. *Dawn*, 26 October 1948, p. 4.
20. US Embassy Karachi to State Department, Washington, Despatch No. 11461, 11 June 1959, 790.00/6-1159, USNA.
21. K.R. Sipe, 'Karachi's Refugee Crisis: the political, economic and social consequences of partition-related migration,' (Duke University PhD, 1976), pp. 158–9.
22. For details of municipal elections in Karachi, see US Embassy Karachi to State Department, Washington, Despatch No. 678, 7 February 1958, 790D.00/2-758, and Despatch No. 1094, 29 May 1958, 790D.00/5-2958, USNA.

23. See discussion of the possibility of the abolition of the provincial system in West Pakistan raised in 'Are Provinces Necessary?', *Daily Gazette* (Karachi), 1 September 1948, and the correspondence which followed between the US Embassy in Karachi and the State Department in Washington, Despatch No. 4405, 9 September 1948, 845D.014/9-948, USNA.
24. Riots between refugees and Pathans, for instance, took place in March 1958 over access to land, water and other facilities, and led to a number of deaths and injuries and a curfew, see US Embassy Karachi to State Department, Washington, Telegram No. 2189, 5 March 1958, 790D.00/3-558, USNA.
25. Akmal Hussain, 'The Karachi Riots of December 1986: Crisis of State and Civil Society in Pakistan', in Veena Das, ed., *Mirrors of Violence: Communities, Riots and Survivors in South Asia* (Delhi, 1990), pp. 185–93.
26. For example, S.N. Eisenstadt, *The Absorption of Immigrants*, (London, 1954) and then later his *The Transformation of Israeli Society* (London, 1985).

8 Illegal Foreign Migrant Workers in Japan: Change and Challenge in Japanese Society

Yoko Sellek

INTRODUCTION

The movement of people is considered to be one of the structural characteristics of the world economy. Capitalist development has always been dependent on the migration of labour both within and across national boundaries. In the context of North–South relations, the industries and transnational enterprises from the highly developed countries of the 'north' expand into the developing countries in the 'south' through direct foreign investment and international subcontracting.[1] Existing industries in the 'south', particularly export-oriented agriculture, manufacturing, and component assembly industries, require a large supply of low-wage workers, and mobilize large numbers of people into wage labour. As a result of labour recruitment, the mobilization of new workers including women into the labour force becomes significant, in turn disrupting traditional work structures.[2] As a result, traditional industries have declined gradually, rural populations have been uprooted and an urban pool of wage labourers and potential migrants has been created. Simultaneously, a foreign presence dominates the new industrial zones objectively and culturally, creating linkages to the countries from which the capital originates.[3]

The social history of industrialization in Japan is no exception to this trend, and the recent movement of illegal migrant workers, involving migrations from the peripheral and semi-peripheral dependencies of Asia to the Japanese centre, reflects not only Japan's regional economic hegemony, but a structural characteristic of the world capitalist economy. This chapter describes the current state of foreign illegal migrant workers in Japan[4] and discusses the reasons why the country has been unable to avoid an influx of such workers. It further describes the ongoing metamorphosis in Japanese society from what has commonly been regarded as a 'homogeneous society' into a heterogeneous, or plural society.

In historical terms, Japan has experienced streams of immigrants and emigrants for a very long time. Although the problem of illegal migrant workers only became serious from the mid-1980s onwards, previously Japan had been one of the major 'exporters' of migrants in Asia. The total number of Japanese emigrants between 1868 and 1945 was approximately 780 000, excluding those leaving the areas of China and Korea that were Japanese colonies during parts of this period and by 1991, the number of Japanese currently living abroad had reached approximately 260 000.[5] The number of Japanese descendants abroad, including those in second and third generations, is estimated to be approximately 1 400 000.[6] Japan has also accepted immigrants. Nearly 700 000 South and North Korean permanent residents are a legacy of the massive migration of Koreans to Japan between 1910 and 1945, and include people who were conscripted for work as replacement labour for the mining and construction industries.

The fact that Japan has experienced both the sending and receiving of migrants has served to influence government policy against contemporary illegal migrant workers. The government perceives that the country is overpopulated to the extent that it has to encourage emigration. It also recognizes that the social, political and economic underpinnings of Japanese society give rise to a highly specific set of beliefs that define the Japanese culture which, through its strength and cohesiveness, provides a fertile breeding ground for discrimination against other Asian nationals.[7] In this sense, the recent problems associated with foreign illegal migrant workers are not new.

Since the mid-1980s the acceptance of foreign workers has become a central issue in Japan. The reason for this is that, although the Immigration Control and Refugee Recognition Act[8] of 1951 provides that the government, except in a very restricted number of cases, may only accept foreign workers who have some special skills or training which cannot be substituted by Japanese nationals, there has been an increasing number of foreigners, especially white-collar workers, who have been attracted by Tokyo's status as a new global financial centre.[9] Furthermore, there is a growing need to make use of foreign labour in Japanese companies. At the same time, the number of illegal migrant workers in the field of so-called unskilled labour who are forbidden from entering the country according to the Immigration Law has risen sharply since the mid 1980s.[10] These illegal migrant workers include those who have engaged in an activity other than that permitted by the status of residence granted them and overstayers continuing in such activities.[11] The influx of illegal migrant workers has engendered a fear in Japan that it might experience the difficulties seen in Western European countries that have received migrant workers, including the unskilled.

TRENDS IN THE NUMBER OF ILLEGAL MIGRANT WORKERS IN JAPAN

The number of foreign residents in Japan is not large compared with that in Western countries. The current figure, including Korean and Chinese permanent residents, is approximately 940 000, comprising only 0.8 per cent of the total Japanese population of more than 120 million. Because neither the government, employers, nor the Japanese people in general anticipated the introduction of a large number of illegal migrant workers, there has been mounting concern about government policy. Since 1987 the fundamental question has been whether Japan should or should not accept unskilled foreign workers and there has been much discussion about the possible ramifications of adopting a more open-door policy towards unskilled workers. While the government has been searching for the best solution, illegal migrant workers have continued to appear in ever increasing numbers in all corners of Japan.

Early Stages of Entry of Illegal Migrant Workers into Japan

The issue of illegal migrant workers in Japan first arose with the influx of female illegal migrant workers from South-East Asia during the 1970s. As Japan's economy began to grow rapidly, organized 'sex tours' for Japanese men to other Asian countries became popular and proliferated until the recipient countries began to discourage such activities at the beginning of the 1980s. In response to the drying up of this lucrative source of income, the same agents who had been responsible for organizing the sex tours to Asian countries began 'importing' Asian women to Japan for similar purposes.[12]

Of the Immigration Law violation cases in the 1970s, each year approximately 250–320 involved unauthorized labour, and 80 per cent of illegal migrant workers were female, originating from Chile, the Philippines and Thailand. After 1979 the number of Taiwanese females who entered Japan as tourists but intending to work increased sharply,[13] and Taiwanese female illegal migrant workers who worked as hostesses in bars or cabaret clubs comprised 80 per cent of all foreigners exposed as working in unauthorized activities in 1979.

At the same time, the numbers of private agents and labour brokers operating in both the sending countries and Japan increased, and an organized and planned movement of foreign workers began to emerge. These private agents and labour-brokers became more ingenious, producing forged passports and visas, providing cheap air tickets, and facilitating the entry into Japan of illegal migrant workers through regional ports where there was less chance of detection. In 1979, out of a total of 667 exposed foreign migrants working in unauthorized activities, 658 entered Japan on sightseeing visas and 70 per cent of the female illegal migrant workers worked in what is known in Japan as the 'water trade' (*mizu shobai*) as hostesses, strippers and prostitutes. Criminal groups were engaged in the recruitment and employment of 93 per cent of female illegal migrant workers.[14]

Trends in the Number and Origin of Illegal Migrant Workers

Table 8.1 shows the trends in the numbers of illegal migrant workers found to be participating in unauthorized activities or residing in violation of visa restrictions between 1984 and 1991 broken down by nationality and sex.

In 1990, owing to the revision of the Immigration Law, a large number of illegal migrant workers decided to declare themselves to the Immigration Bureau – this led to almost a doubling of their number compared with 1989. In 1991 nearly 33 000 foreign nationals worked illegally, a 10 per cent increase on 1990 and approximately a 140 per cent increase on the figure for 1983.

This sharp increase in illegal migrant workers since the mid-1980s results mainly from the rapid increase in the number of male illegal migrant workers. Until 1985 more than 90 per cent of illegal migrant workers exposed were female. Since 1985, however, the number of female illegal migrant workers has not changed significantly. In 1988, the number of male illegal workers surpassed the number of females for the first time and, in 1991, 77 per cent of illegal migrant workers were male.

Table 8.1 underlines significant features of the national origins of migrants.

1. Most illegal migrant workers in Japan are from developing countries in Asia. Koreans, Malaysians, Filipinos, Thais, Chinese (including Taiwanese), Pakistanis and Bangladeshis comprise more than 95 per cent of those exposed up to 1989; this figure is a little over 70 per cent for 1991.

… broken down by nationality and sex, 1984–91

Nationality/ Region		*1984*	*1985*	*1986*	*1987*	*1988*	*1989*	*1990*	*1991*
Total number		4783 (350)	5629 (687)	8131 (2186)	11307 (4289)	14314 (8929)	16608 (11791)	29884 (24176)	32908 (25350)
South Korean		61 (34)	76 (35)	119 (69)	208 (109)	1033 (769)	3129 (2209)	5534 (4417)	9782 (8283)
Iranian		–	–	–	–	–	15 (13)	652 (648)	7700 (7611)
Malaysian		–	–	–	18 (15)	279 (265)	1865 (1691)	4465 (3856)	4855 (3892)
Thai		1132 (54)	1073 (120)	990 (164)	1067 (290)	1388 (369)	1144 (369)	1450 (661)	3249 (926)
Filipino		2983 (96)	3927 (349)	6297 (1500)	8027 (2253)	5386 (1688)	3740 (1289)	4042 (1593)	2983 (1079)
Chinese:	Mainland					7 (5)	39 (26)	481 (428)	1162 (981)
	Taiwan	466 (136)	427 (125)	356 (161)	494 (210)	492 (223)	531 (275)	639 (351)	460 (225)
	Hong Kong					3 (2)	18 (15)	22 (20)	43 (36)
Pakistani		3 (3)	36 (36)	196 (196)	905 (905)	2497 (2495)	3170 (3168)	3886 (3880)	793 (793)
Sri Lankan		–	–	–	–	20 (20)	90 (87)	831 (821)	307 (295)
Bangladeshi		0 (0)	1 (1)	58 (58)	438 (437)	2942 (2939)	2277 (2275)	5925 (5915)	293 (292)
Other		138 (27)	89 (20)	115 (38)	150 (70)	267 (154)	590 (374)	1957 (1586)	1281 (937)

Notes: 1. Figures in brackets indicate the number of males included in the total.
2. Dashes indicate that a separate figure for the number of nationals is not available. In such cases these nationals are included in the figure for the category entitled 'Other'.

Sources: Immigration Control Office, Ministry of Justice.

2. A large number of female illegal migrant workers are from the Philippines and Thailand, whereas, in the case of male illegal migrant workers, the countries of origin are becoming far more diverse.[15]
3. Since 1988 the numbers of Korean and Malaysian illegal migrant workers have increased significantly. For the Koreans, the reasons behind this trend are the Korean government's relaxation on exit visas in 1989 and the consequent impetus to Korean migration. Such migrants used the existing and well-established network of Japanese-Korean permanent residents in Japan.[16] A further factor was that the revised Immigration Law aimed specifically to enforce the penal codes against employers who employed illegal migrant workers entering Japan on or after 1 June 1990. This had the effect of encouraging employers to take on illegal migrant workers, such as Koreans and Chinese-Malaysians, who have a similar physical appearance to those of native Japanese, to minimize the possibility of discovery. As long as such employees do not show their non-native proficiency in the Japanese language, their chances of arrest are generally low.
4. The existence of visa-exempt agreements between the Japanese government and other countries is closely related to the influx of illegal migrant workers. In 1988 when much attention was focused on the issue of illegal migrant workers, nearly two-thirds of all such male workers were from Pakistan and Bangladesh. At that time, Japan had visa-exempt agreements with these two countries and, with no visa requirement for tourist visits of up to three months, many Pakistanis and Bangladeshis entered Japan as tourists and stayed on to work. However, to cope with the influx of illegal migrant workers from these countries, Japan decided to suspend the visa-exempt agreements in 1989, which subsequently led to significant decreases in the numbers of Bangladeshi and Pakistani illegal migrant workers in 1991. Thereafter only Singapore, Brunei and Malaysia, among Asian countries, retained such agreements with Japan. The acceleration in the influx of Malaysian illegal migrant workers is due largely to the existence of a visa-exempt agreement.

Japan also entered into a visa-exempt agreement with Iran from 1974, shortly after the first oil crisis.[17] Owing to the decline of domestic economic conditions in Iran, many Iranians sought work in the oil-producing countries of the Middle East.[18] However, with the outbreak of the Gulf War in August 1990, most of these migrant

> workers had to evacuate to neighbouring countries, and economic opportunities in the region quickly dried up.[19] Around the same time, the number of Iranian illegal migrant workers in Japan began to increase rapidly, and 7700 Iranians were working illegally in Japan in 1991, representing about 12 times the figure of 652 for 1990. The number of Iranian illegal migrant workers is, however, expected to decrease rapidly in the near future following the Japanese government's suspension in April 1992 of its visa-exempt agreement with Iran for short-term visitors.

The numbers of exposed illegal migrant workers shown in Table 8.1 represent only the tip of the iceberg in relation to actual numbers. According to a report by the Justice Ministry's Immigration Bureau released in June 1992, a total of 292 791 foreign nationals were in Japan illegally as of 1 November 1992, representing an increase of 13 899 on the number six months earlier.[20] Among these, 242 679 entered on short-term visas and accounted for 82.9 per cent of all foreigners staying in Japan illegally. The report also noted that 18 112 foreigners staying in Japan illegally entered the country on pre-college student visas, and 6296 entered on entertainment visas.[21]

In addition to overstayers, numerous would-be workers are detected attempting to disguise their purposes of entry, for example, through the use of forged passports or visas, and are refused entry. In 1987, only 4151 persons were refused entry into Japan, but, in 1991, this figure had risen to 20 726, more than double the 9669 who were turned back in 1990.[22]

The category of 'pre-college student' has proved convenient for foreigners coming to Japan with the intention of working. Most foreigners on pre-college student visas enter either Japanese language schools or business schools and usually intend to study the language in order to enter Japanese universities or to acquire some skills in industrial or commercial fields. The influx of pre-college students has become substantial since 1983, and the beginning of this influx coincided with a proposal made by the former Prime Minister Nakasone in 1983 to accept 100 000 foreign students by the beginning of the twenty-first century as part of a campaign towards the internationalization of Japan. Consequently, since 1984 the government has significantly simplified the procedure for the issue of pre-college student visas. For example, the head of the Japanese language school which the foreign pre-college student attends can now act as guarantor for the student. It has also become easier to extend the period of stay permitted under such visas. This relaxation produced the ironical result that the number of pre-college students intending to work in Japan increased. Table 8.2 compares the recent number of foreigners entering Japan on pre-college student visas with the number of foreign students awarded scholarships to study at Japanese universities.

The number of foreigners entering Japan on pre-college student visas was 4140 in 1984, but by 1988 this figure had risen eightfold to 35 107. Pre-college students from Asian countries constitute roughly 90 per cent of the total for this category.

Foreigners studying at Japanese language and business schools in Japan are permitted to work a maximum of four hours a day if they report this fact to the Immigration Bureau.[23] In reality, however, pre-college students do not observe these regulations. Part-time work earns them insufficient wages to live even modestly and pay tuition fees. They are faced with little choice but to work longer hours without reporting this extra work to the authorities.

According to a survey undertaken by the *Peking Daily News* in 1988 of about 100 Chinese pre-college students in Japan, 30 per cent had come to Japan intending to earn money, 30 per cent had come intending to enter higher education as well as earn money, and 20 per cent had no specific reasons other than the desire to study abroad.[24] Another report indicates that 80–90 per cent of pre-college students in Japan are there specifically to

Table 8.2 Trends in the number of foreigners entering Japan on pre-college student visas and on foreign student visas, 1982–91

Year	*Pre-college students*	*Foreign students*
1982	2 556	3 410
1983	3 448	3 912
1984	4 140	4 329
1985	8 942	4 997
1986	12 637	5 419
1987	13 915	5 812
1988	35 107	6 435
1989	18 183	7 777
1990	20 851	9 528
1991	20 654	9 620

Source: Ministry of Justice.[a]

[a] The figures given for pre-college students and foreign students between 1982 and 1990 are extracted from Tanaka (1991), p. 175; those for 1991 are extracted from data published by the Immigration Bureau, Ministry of Justice (1992), p. 105.

earn money, for example, for consumer purchases or to start small businesses. There are numerous similar reports and it is difficult to obtain reliable figures and draw general conclusions. However, since living expenses, schools fees, and accommodation costs are extremely high in Japan, and the Chinese government does not permit its citizens to take out more than 10 000 yen (approximately £58) from China, it would seem reasonable to assume that most Chinese pre-college students have to take part-time jobs. According to an investigation carried out by the Council for Pre-College Students, 65 per cent of these students were working in part-time jobs, 49 per cent in restaurants, and 22 per cent involved in other service industries, operating as cashiers in supermarkets, newspaper delivery boys/girls, or clerks.[25]

There are many unscrupulous language schools acting as private employment agencies with support from brokers, and these form a network for pre-college students working illegally. In conjunction with the increasing number of pre-college students, the number of Japanese language schools increased significantly until 1988. Although there are today many reputable language schools based on the Regulations of School Education in Japan, initially there were neither guidelines nor permission given by any of the government ministries in Japan, including the Ministry of Education, for the establishment of such schools. It became common for people working in, for example, estate, accommodation, and travel agencies, and in preparatory schools, to establish language schools with the intention of making quick and handsome profits.

There were only 49 Japanese language schools in 1984, but this number continued to increase until 1988 when there were 309.[26] The ways in which they have exploited pre-college students include, for example, selling the certificates which permit these students to enter the schools, and charging fees for introducing them to the guarantors who are required in order for the students to apply for their visas, for introducing students to landlords and employers able to offer part-time jobs, for falsifying attendance records, and awarding false marks in the examinations that they must pass in order to extend their stays. The principal functions of the brokers are travel arrangements for foreigners who wish to work in Japan, fabricating the bank statements and other documents required for visas, and placing students in part-time jobs. Pre-college students usually have to pay the brokers more than 300 000 yen (approximately £2000) a sizeable sum for most Chinese.[27]

As can be seen from Table 8.2, the number of pre-college students increased rapidly until 1988, but 1989 saw a substantial decrease. This decrease resulted from a decision by the government to impose a restriction on the issue of pre-college student visas to Chinese applicants. Furthermore, in June 1989, the Association to Promote Japanese Language Education was established and an inspection and authorization programme for Japanese language schools began. In February 1991, 449 Japanese schools were members of the Association, a high level of affiliation mainly because one pre-condition for a pre-college student visa for study at a particular school is that the school must belong to the Association. Currently, owing to the decrease in the number of pre-college students, many Japanese language schools are finding it difficult to remain in business. It is estimated, however, that at least 35 000 pre-college students are working illegally in Japan[28] and there is a growing trend for Japanese language schools to try alternative strategies for staying in business by using the system, for example, to increase their numbers of trainees.[29]

The diversity of types of occupation in which exposed illegal migrant workers are involved has increased recently. Table 8.3 shows the trends between 1988 and 1990, broken down by the type of job undertaken. Until 1989 the largest number of male illegal migrant workers was found in the construction industry, followed by the manufacturing industry.[30] In 1990, however, the number of male illegal migrant workers in manufacturing surpassed the number in construction for the first time.[31] Of female illegal migrant workers, more than 80 per cent of those exposed worked as hostesses in 1988, while in 1990 this figure had decreased to 55.6 per cent. By contrast, in the manufacturing industry the number of exposed illegal female workers increased in 1990 to over 13 per cent from only 3.1 per cent in 1988.[32] Thus, in the cases of both male and female illegal migrant workers, more workers have shifted their jobs to the manufacturing, commercial and service industries.[33]

EXPLANATIONS FOR SIGNIFICANT INCREASES IN THE NUMBER OF ILLEGAL MIGRANT WORKERS SINCE THE MID-1980s

Economic migration is governed by various incentives, but these depend on the value-judgements of each individual migrant. It is widely believed that the principal factors behind the migration process are the existence of a differential in income between that of the sending and receiving regions and the employment opportunities at the other end of the migration stream.[34] Further, both sending and receiving countries think of the migration as temporary. In the sending country, the migration is supposed to relieve population pressure and help reduce unemployment, to generate scarce foreign exchange, and help in developing the skills

Table 8.3 Trends in the numbers of exposed migrant worker broken down by type of job undertaken, 1988–90 (percentages in brackets)

Male migrant workers

Type of job	*1988*	*1989*	*1990*
Construction	3 807 (42.6)	5 581 (47.3)	6 775 (40.2)
Manufacturing	3 486 (39.0)	4 696 (39.8)	7 617 (45.2)
Unskilled services	765 (8.6)	575 (4.8)	843 (5.0)
Shopkeeper	283 (3.2)	227 (1.9)	253 (1.5)
Cook	115 (1.3)	121 (0.9)	253 (1.5)
Others	473 (5.3)	591 (4.9)	1 111 (6.6)
Total (100)	8 929 (100)	11 791 (100)	16 852 (100)

Note: For 1990 the figures are based on the 16 852 male migrant workers whose jobs were revealed out of a total of 24 176 exposed male migrant workers.
Source: Ministry of Justice

Female migrant workers

Type of job	*1988*	*1989*	*1990*
Hostess	4 704 (87.3)	3 559 (73.8)	2 605 (55.6)
Manufacturing	165 (3.1)	323 (6.7)	614 (13.1)
Unskilled services	120 (2.2)	276 (5.7)	374 (8.0)
Shopkeeper	105 (1.9)	227 (4.7)	46 (1.0)
Others	291 (5.4)	432 (8.9)	1 046 (22.3)
Total (100)	5 385 (100)	4 817 (100)	4 685 (100)

Note: For 1990 the figures are based on the 4685 female migrant workers whose jobs were revealed out of a total of 5708 exposed female migrant workers.
Source: Ministry of Justice.

necessary for certain emergent industrial sectors. On the other hand, in the receiving country the migrants are considered to complement the indigenous labour force since they take jobs which indigenous workers no longer wish to undertake, and in this way serious labour shortages can be overcome.[35]

In the case of Japan there are three principal considerations influencing the decision to migrate: first, disparities in economic development among countries, particularly in terms of income levels; second, imbalances between supply and demand of labour among countries; and third, the existence of institutional mechanisms or networks that promote international movements of labour.[36] It is reasonable to conclude, therefore, that the coexistence of these three factors explains the large surge in illegal migrant workers to Japan since the mid-1980s. These factors have functioned to 'push' the migrant workers out of their countries of origin and 'pull' them into Japan.

Disparity in Economic Development – Economic Incentives

The main attraction of Japan for illegal migrant workers is that economic might has pushed income levels far above those in neighbouring countries. The sudden appreciation of the yen after the Plaza Agreement in 1985 served to magnify this gap. The corresponding wage differentials mean that even employed workers in the developing countries of Asia are attracted to work in Japan. In 1988 when the problem of illegal migrant workers escalated, the per-capita income of Japan was 123 times that of Bangladesh, 64 times that of China, 60 times that of Pakistan, and 33 times that of the Philippines.[37] So, for a Bangladeshi, one month's wages in Japan is equivalent to 32 months of work in his native country, even if for his work in Japan he is paid only one half of the amount a Japanese worker would expect to earn in a similar job. In practice, however, the high cost of living and the double structure of income between large and small companies and between full and part-time employment diminish those differentials. Similarly, the migrant worker may well not be paid in full because of possible illegal 'kickbacks' the employer might demand, non-payment of wages, or dismissal. In countries with much lower income levels, however, the streets of Japan do undoubtedly appear to be paved with gold.

Incentives Generated by the Imbalance between Supply and Demand of Labour among Countries

Pay differentials, however, are not the only important incentive to migration, as income levels have been consistently higher in Japan than in the rest of Asia since the late 1960s, but immigration remained sluggish until many years later. The recent increase in the number of illegal migrant workers in Japan results primarily from a rapid increase in demand for unskilled workers in the Japanese labour market. But why should employers choose to take on illegal migrant workers in full knowledge of the fact that such employment is in blatant contravention of the law?

From the mid-1980s until very recently Japan has had a severe shortage of labour. Although labour shortages have been a recurrent problem since the high-growth years of the 1960s when the possibility of importing labour was first considered, the shortages were overcome through reliance on seasonal migrations from rural areas within Japan.[38] By the end of the 1960s the number of seasonal workers available for industrial labour had fallen off sharply.[39] Despite discussions about the possible introduction of foreign migrant workers in unskilled industrial fields, the government decided not to accept such workers in 1967, 1973 and 1976.[40] According to Iyotani, the time-lag between the 1960s, when the labour shortage in Japan became apparent, and the 1980s, when the number of foreign illegal migrant workers began to increase significantly, resulted from the following factors:

1. The Japanese government managed to raise labour productivity by improving manufacturing techniques, and consequently demand from employers for foreign-migrant labour was not strong. In the 1970s, further technological improvements in manufacturing processes were achieved through developing the technology of micro-electronics, mechanization and automation, and as a result there was less demand for unskilled workers in the labour market;
2. Long working hours were maintained, although these decreased from 2660 hours annually in the mid-1960s to 2100 hours in 1970. Also, women and elderly people were encouraged to work, thus contributing to the maintenance of the supply of low-wage labour;
3. Appropriate conditions were provided to maintain the flexibility of the labour market, such as technological innovation, the flexible hiring policies of companies, and the supply of temporary and part-time workers;
4. Following the oil shocks of the 1970s, economic growth in Japan slowed down; and
5. There was an increase in direct investment from Japan into South-East and East Asia where cheap and ready sources of labour existed.[41]

After the oil crises, the labour force expanded continuously by more than 1 per cent per year and was sufficient to match Japan's economic growth rates through the mid-1980s. In the mid-1980s, however, Japan's economy began to grow rapidly and from 1987, in particular, the shortage of labour became an extremely serious problem. The seasonally-adjusted active opening rate (the number of job openings divided by the number of applicants for those jobs) was stable at around 0.6 from 1975 to 1986, implying that six jobs were available for every 10 applicants. By 1991 the rate had reached 1.4, meaning that 14 jobs were now available for every 10 applicants. This was the highest rate since the first oil crisis in 1973.[42] The shortage of young urban labour has become especially serious.[43]

Since the collapse of the speculative 'bubble economy' in 1991, Japan has entered a period of economic slowdown. A recent government survey shows that the excessively tight labour market has been easing, and the seasonally-adjusted active opening rate has been declining, and in 1992 was down to 1.14.[44] The main growth industries in Japan's economy, such as automobiles and electronics, are facing a sharp fall in demand, and the service sector is facing its first recession.[45] The shortage of labour has, however, continued in the small business sector, the manufacturing industry, the wholesale and retail industries, and in the labour-intensive service industry.[46]

Based on demographic trends, it is estimated that Japan will face a structural, and not a cyclical, shortage of labour in the near future. After the anticipated labour peak in the mid-1990s, the number of school-leavers will begin to decrease (corresponding to a general decrease in the number of young people) and will not increase again for at least two decades.[47] In 1990, life expectancies were 76 and 81 for men and women respectively, the highest in the world, and the average birth rate had plummeted to 1.53.[48] As the population ages, young people make up a progressively smaller fraction of the labour force. The general shortage of labour will continue and it is estimated that the annual rate of growth of the labour force will fall to 0.8 per cent in the period 1989–95 and then further to 0.4 per cent in the period 1995–2000, while the unemployment rate will be 2 per cent of the total population by 1995, declining to 1.8 per cent by the year 2000.[49]

i. Construction Industry

The construction industry was thriving during the mid-1960s, and was especially buoyant just before the 1964 Tokyo Olympics. Many workers joined the industry during that period, including a large number of seasonal workers from farming areas. However, when the government implemented an 'administrative reform' in the early 1980s to restrain public as well as private-sector investment in plant and equipment, a recession occurred and most of the industry's workers left their jobs. Moreover, very few young workers joined the industry's workforce during this period.

Since 1986 the construction business has been picking up continuously as a result of government policies to stimulate domestic demand and to ease restrictions on development while pushing ahead with public-spending projects in co-operation with the private sector. The sudden increase in funds in real estate provided by banks helped accelerate the redevelopment of cities and encouraged the construction of new buildings. The total amount of annual investment in the construction industry was around 50bn yen between 1980 and 1985, but increased to 54bn yen in 1986, 61bn in 1987, 66bn in 1988, and was over 70bn yen in 1989.[50]

The construction industry, however, did not have a sufficient stock of labour to cope with the boom. There was a serious shortage of skilled and unskilled workers and, in addition, about 40 per cent of workers employed in construction in 1988 were over 50 years old. The main reason for this shortage is that labour conditions in the construction industry are generally considered to be far worse than in other industries.[51]

The expression '*Roku-K*' (six-K) in Japanese is commonly used to describe such industries as being '*Kitanai* (dirty)', '*Kitsui* (hard-work)', '*Kiken* (dangerous)', '*Kyuryo ga yashui* (low-wages)', '*Kyujitsu ga sukunai* (few-holidays)', and '*Kakko ga warui* (not-smart-in-appearance)', and conveys vividly the unpopularity of the industry.[52]

ii. Manufacturing Industry

To minimize the cost of labour, the larger manufacturing companies try to increase their investment in countries where labour costs are low and to update their technology using, for example, flexible production systems and computer-integrated manufacturing systems. The shortage of labour in Japan is however more serious among the smaller manufacturers that have played such a major role in the country's economic development during the post-war period. The situation within these companies is very similar to that in the construction industry, particularly the subcontracting system. A strong relationship usually links parent companies to each of their subcontractors, and recently parent companies have begun to exercise more rigid control over their subcontractors by, for example, demanding that they keep down the costs of goods and shortening time limits for delivery, paralleling the 'just-in-time' system of the automobile industry. Under such circumstances a flexible employment system is required involving, for example, night-shifts or three lengthened working days a week. Good working conditions are therefore often neglected, resulting in a serious shortage of labour combined with a trend towards an older workforce.[53]

iii. Service Industry

Specialization has been increasing in Japan, based on the mass-production system that has underpinned the country's economic development since World War II. In particular, since the 1970s Japan has been transformed into a 'soft-industry society' in which an advanced service industry constitutes a major component. As one of the financial centres of the world, highly capital-intensive industries such as banking and finance have proliferated in Tokyo, creating a pool of highly-paid professional, technical, managerial and administrative jobs. New types of service industry have emerged, such as the information, the manpower supply, and the building maintenance industry, with each spanning traditional industrial boundaries. Simultaneously, however, their development has generated demand for miscellaneous low-wage jobs and, in general, there exists a shortage of labour in those fields where working conditions are poor, and mechanization is impossible to implement.

iv. Sex-Related Industry

In Japan the sex-related industry is known as 'the 10bn-yen industry'. The job opportunities in this industry are numerous and strong demand for workers has always existed for several reasons. Young, attractive women are always in strong demand and older workers are usually abandoned. The turnover rate is therefore high. A tolerant attitude towards prostitution has long existed in Japanese society and still exists today. It stems from the still commonplace discrimination against women in Japan and the household system in which the husband is the head. Women have traditionally been considered by Japanese men to fall into two different functional categories: those who produce children for households, and those who provide sexual enjoyment. Neither category has been treated with dignity. Even today such discrimination still exists in Japanese society, and there is a tacit acceptance by society of men's involvement in extra-marital affairs. Moreover, it has almost become fashionable for unmarried women to be involved in such affairs. Japanese businessmen are 'soldiers of companies' who are forced to work long and hard and to submit to control by their companies. Such men, who often experience feelings of isolation, partake of the favours provided by the sex industry allegedly to dispel their gloom. In addition, a large number of businessmen use bars to entertain their clients. Because the industry is growing and profitable, prostitution particularly is an extremely time and cost effective way for employers to make money. It is estimated that the employers of prostitutes can earn 30 000 to 40 000 yen (approximately £175–235) per client. From this income they might pay each prostitute only 8000 yen (about £50) or less, and sometimes even nothing.[54]

RESPONSES TO THE INCREASE IN THE NUMBER OF ILLEGAL MIGRANT WORKERS

Government

1. Revision of the Immigration Law

In order to control the significant increase in the number of illegal migrant workers of the mid-1980s, the Ministry of Justice proposed a revision to the Immigration Law. This revision was accepted by the Cabinet for submission to the National Diet in March 1989 and enforced with effect from 1 June 1990. The purpose of the revised Immigration Law was to respond to the growing need for foreign labour and to strengthen the measures to control illegal migrant workers.[55] The revised Immigration Law made the following provisions:

(i) In order to respond to the needs of Japanese companies, the scope of working activities for foreigners was enlarged in certain skilled and professional categories. New types of status of residence cover: (a) specialists in human science fields, such as translators and interpreters; (b) those engaged in legal and accounting services, such as lawyers and accountants; (c) those engaged in medical services, such as doctors and dentists; (d) research workers; (e) foreign language teachers; (f) intra-company transferees; (g) cultural activists; (h) pre-college students, such as Japanese language school students as opposed to university students; and (i) those offering specialist skills, such as software engineers.[56] The revised law has also expanded the existing categories of status of residence, and has rationalized the total of 28 residence-permit categories, including those unrelated to the purpose of employment and those with no special restrictions on activity, into six broader categories in an attempt to make the law unambiguous for foreigners residing in Japan.[57] In addition, it has simplified and expedited immigration procedures by clearly identifying the documentation required for particular categories of status of residence.[58]

(ii) In order to strengthen the measures against illegal foreign workers, the revised law enforces penal codes against those employing illegal migrant workers who entered Japan on or after 1 June 1990, and

> against brokers who have found jobs for illegal migrant workers who have entered Japan since then. These employers and brokers are fined up to a maximum of two million yen (about £11 700), or are imprisoned for up to three years.[59] Also, the revised law has clarified the status of residence for foreigners to make it easier for potential employers to determine whether a particular foreigner may be employed legally. At the same time, the law provides a foreigner authorized to work with an employment certificate, in order to avoid confusion on the part of employers regarding the eligibility of a foreigner for employment.[60]

Owing to the further tightening of the regulations relating to the acceptance of illegal migrant workers under the revised Immigration Law, large numbers of employers from small and medium-sized companies, suffering seriously from shortages of labour, lodged complaints with the government to the effect that they could be forced to close down their businesses since there would be no labour force available to them. As a result, the government sought to create other means of providing a low-cost source of labour. The approach it adopted was to expand acceptance of foreign trainees and to promote the employment of South American/Japanese descendants.

2. Foreign trainees

Workers categorized as 'trainees' are considered very useful in alleviating labour shortages. Trainees are 'supposed to conduct activities to learn and acquire technology, skills or knowledge in public or private organizations in Japan'.[61] The government position is that these people are not paid workers, and employers must not permit them to work. However, among the various types of designated training is so-called 'on-the-job training', and if employers adopt this type of training, their 'trainees' are in effect permitted to work. According to the 'Admissible Standards for Foreign Trainees',[62] one-third of this training is to be carried out in classrooms, but the number of cases where the employers are ignoring this ordinance and increasing the amount of on-the-job training has risen.[63] The Ministry of Justice has deemed these cases to be 'working under the pretext of training' and has tried to conduct careful surveillance. The Ministry further established the 'Prior Examination Standards for the Landing of Foreign Trainees' in 1990 in response to the change in the concept of a 'trainee'. In practice, however, the government cannot keep pace with the increasing number of trainees.

Table 8.4 shows the number of trainees according to country of origin for the years 1982 to 1991. The number of foreigners entering Japan on

Table 8.4 Trend in number of trainees broken down by nationality, 1982–91

	1982	*1983*	*1984*	*1985*	*1986*	*1987*	*1988*	*1989*	*1990*	*1991*
Chinese	1 217	1 495	2 122	1 541	2 848	2 688	3 840	3 496	7 624	10 668
Thai	670	899	1 035	1 233	950	2 428	4 708	4 502	5 075	6 290
Korean	829	1 366	1 505	1 604	2 336	2 800	3 343	4 125	4 485	4 439
Malaysian	430	867	753	610	620	757	1 329	2 175	3 564	4 307
Philippine	794	685	927	1 134	968	1 207	2 464	3 974	3 460	4 476
Indonesian	–	–	1 109	1 135	1 114	1 310	1 378	1 748	2 891	3 883
Others	6 033	6 617	5 811	6 730	5 552	5 891	6 370	9 469	10 467	9 586
Total	9 973	11 929	13 262	13 987	14 388	17 081	23 432	29 489	37 566	43 649

Source: Ministry of Justice.[b]

[b] The figures given for the number of trainees between 1982 and 1990 are extracted from Tanaka (1991), p. 208; those for 1992 are extracted from data published by the Immigration Bureau, Ministry of Justice (1992), p. 105.

trainee visas in 1987 was 17 081. In 1990, this number had increased to 37 566 and in 1991 was 43 649.[64]

The largest number of trainees in 1990 was from China, more than double the previous year. Chinese who had intended to enter Japan on pre-college student visas were unable to do so as a result of the tightening on the issue of this type of visa; instead these Chinese have come on trainee visas.

In May 1992, the Ministry of Labour introduced a proposal entitled 'The New Two-Year Trainee Plan' with the aim of accepting up to 100 000 trainees for two-year stays to include 15 months of work. This proposal is designed to address the problem of the rapidly increasing numbers of illegal migrant workers, to help alleviate further the labour shortage and, for the first time, to permit unskilled workers to enter Japan to acquire specific skills. Training is to be provided in two stages: three months of lectures and other off-the-job courses, and six months of actual skills training to be followed by a technical skill evaluation; some trainees will be allowed to work for up to 15 months in on-the-job training. During this period, the trainees will be covered by all labour-related laws, including the Labour Standards Law.[65] The new trainees are to be accepted through the Japan International Training Cooperation Organization, which was established in 1991 under the control of the five ministries of Labour, Justice, International Trade and Industry, Foreign Affairs, and Construction.[66]

It is possible that such trainees will find themselves open to exploitation. Most receive extremely low allowances compared with illegal migrant workers because the government stance is that allowances represent 'pocket money' and not wages.[67] Government guidelines indicating that the allowance must be low actually provide a ready excuse for unscrupulous employers who wish to use trainees as unskilled workers on low remuneration. Also, at present, labour insurance, such as employment, health, welfare, pension and workers' accident compensation insurance do not apply to trainees. Of these, ineligibility for workers' accident compensation insurance is probably the most serious problem since, if a trainee is injured or becomes ill in the course of work, he or she is dependent on national health or on private company insurance, the premiums for both of which are frequently neglected by employers despite their legal obligation to provide accident cover.

Contracts also often require that trainees must not disobey the orders of their organizers or give up their training during the course. Should training be abandoned, this will automatically lead to deportation. If the organizations employing the trainees are Japanese companies, it is common for the trainees to be required to work for these companies in their home countries for a certain period of time after their return.[68]

3. The South American-Japanese influx

The number of companies employing South American-Japanese descendants who are *legally entitled* to work in the unskilled sectors has been increasing since 1989. This is because, in conjunction with the revision of the Immigration Law, formal status of residence has been granted to South American-Japanese descendants. From around 1989, companies started to look for foreigners who were permitted to work legally in Japan. The first generation of Japanese descendants has Japanese nationality and has no problem working in Japan: they have simply returned home. Members of the second generation, who do not have Japanese nationality, have been given the status of residence of 'spouse or child of Japanese national', and those in the third generation have been given a status of 'long-term resident'.[69] The spouses of these second and third generation members have also been given the status of 'long-term resident'.[70] They usually enter Japan on sightseeing visas and later change to the status of 'spouse or child of Japanese national', or 'long-term resident'. There are no restrictions on the activities that these individuals can undertake in Japan, so they are able to work legally in the unskilled job sectors. For employers, the benefit of taking on these South American-Japanese descendants is, of course, that they can be employed openly, although the wages commanded are higher than those for illegal migrant workers.[71] For the descendants, working in Japan is very attractive, since the average wage in Japan is, for example, roughly 10 times that in Brazil.

Such descendants are mainly Brazilian, Argentinian and Peruvian, with Japanese-Brazilian descendants being particularly numerous. The Ministry of Foreign Affairs estimates that in 1988 the number of South American-Japanese descendants in Japan was approximately 8500. But by 1991, the total number had reached 148 700, with Brazilians comprising 120 000 and Peruvians 18 000.[72]

The numbers of both foreign trainees and South American-Japanese descendants increased during the period when the government was strengthening its controls over illegal migrant workers. Approximately 20 000 illegal migrant workers have probably returned to their home countries because of the enforcement of the revised Immigration Law, perhaps to be replaced by an influx of trainees and South American-Japanese descendants.

Industrial Sector

Generally the medium and small-sized companies are the principal employers of illegal migrant workers. In the early stages of labour migration into Japan, there were many cases in which illegal migrant workers undertook ancillary roles, doing relatively light or assistant-type work, while Japanese workers filled more central roles.

In conjunction with the expansion of sales, production also expanded, but soon, once employers could not employ a sufficient number of Japanese workers to take full advantage of the expanding economy, it became necessary for illegal migrant workers to undertake more important jobs. Inevitably, the percentage of illegal migrant workers in factories has risen to the point where some have an entirely foreign production staff. Although the highest concentration of illegal foreign workers are in and around the three big city areas of Tokyo, Osaka and Nagoya, these workers are now also found away from the major cities in, for example, the Tohoku district in northern Honshu. Their labour is in high demand for subcontracted work in the automobile, machinery, construction, electronics and agricultural industries as well as in the service industry, for example, in hospitals. These industries are at the heart of the Japanese economy and society; thus regardless of the preferences of government and people, it will be extremely difficult to maintain the country's current economic and welfare systems unless Japan accepts foreign workers and acknowledges their critical contribution to the Japanese labour market.

Since 1988 the industrial sector has begun to express its opinions on the issue of illegal migrant workers. In that year the Japanese Federation of Employers Association submitted a report which stated clear objections to the acceptance of unskilled illegal migrant workers, primarily because most of the members of the federation traditionally belong to the very large heavy-industrial sectors such as steel, chemicals and shipbuilding and hence tend not to suffer from chronic shortages of labour. Since 1989, however, opinions voiced by the industrial sector have moved quickly towards supporting the acceptance of illegal migrant workers, and at the beginning of 1989 proposals favouring this acceptance began to be submitted in rapid succession. Organizations such as the Kansai Committee for Economic Development, the Japan Committee for Economic Development, and the Tokyo Chamber of Commerce and Industry proposed a system under which unskilled workers could be accepted.

In 1990, the Kansai Committee suggested a 'Manpower Centre' to employ foreign workers, including unskilled workers (that is, illegal migrant workers) on a monopoly basis in order to place them with Japanese companies. Under this plan, the Centre would also check whether the foreign workers were being treated at least equally in comparison with indigenous Japanese workers, pay their airfares to and from Japan, and take responsibility for providing facilities for their education, accommodation, and general welfare.

The Kansai Committee further proposed a plan to secure bilateral agreements between Japan and each sending country, covering the kinds of jobs and numbers of workers to be accepted under the scheme. The foreign workers would receive a contract from the Manpower Centre ensuring conformity with Centre regulations for employment. Since the Centre would operate as a monopoly, companies attempting to employ foreign workers by other means would be severely penalized. The Committee also expressed sharp criticism of the attitude of the government, saying that it was only trying to postpone reaching a conclusion and, in so doing, tacitly supported the current situation.[73]

These and other proposals reflected general concerns about the prevailing labour situation in Japan at the time: although the country was enjoying a business boom, the severe shortage of labour was suddenly becoming very serious, as shown by the number of vacant jobs. Since the longer-term prospects are that shortages of labour will become worse, the industrial bodies emphasized the need to accept foreign workers in order to guarantee an adequate labour force in the future.[74] With regard to illegal migrant workers, the proposals clearly stated that the government should pay the social costs for these workers, and should treat them on an equal basis with Japanese workers. Furthermore, the industrial bodies proposed that there should be two forms of acceptance of illegal migrant workers: one being to accept them as workers, and the other to accept them as trainees whose aims are to acquire certain skills in the course of their work. This latter form of acceptance clearly influenced the government in establishing the new two-year trainee plan described above.

INFRINGEMENTS OF HUMAN RIGHTS

In parallel with the increase in the number of illegal migrant workers in Japan, infringements of human rights are becoming more commonplace since migrant workers do not have legal status and so enforcement of their labour rights is precluded. Criminal gangs take advantage of their illegality and traffic in illegal workers. Hard labour under virtual confinement, forced prostitution, work without pay, non-payment of wage arrears, reduction in wages without explanation, dismissal without proper notice, exploitation by brokers, and lack of compensation for injuries due to work-related accidents, are prevalent.[75]

Japan has an extensive network of labour laws designed to protect labourers from exploitation[76] and all these are applicable to foreign workers regardless of their status. However, if illegal migrant workers were to seek protection under the labour laws, they would have to come out into the open and their illegal status would be discovered. For most, exposure and subsequent forced deportation clearly does not outweigh the benefits of being liberated from their 'illegal' maltreatment. Furthermore, in October 1990, the Health and Welfare Ministry issued a notice to illegal residents that they could not qualify for the public welfare scheme.[77] Since April 1992, foreigners with the visa status of 'short-term stay' have been ineligible for cover under the national medical insurance scheme, so that if they are injured or become ill they have to pay the full amount of any medical bills incurred.[78] In the present situation most workers choose to grin and bear their situation, even if their lives are extremely miserable.

Nearly 50 civilian-based organizations trying to support foreign illegal workers who suffer infringement problems had been created by the beginning of 1992. These organizations have criticized the present Immigration Law for not fully covering the rights of foreign workers, and have given emergency relief to foreign illegal workers who have encountered various forms of discrimination.[79]

JAPANESE SOCIETY

According to the Ministry of Justice, the number of Asians who came to Japan in 1991 was 2.5 million, representing roughly 65 per cent of the total number of foreign nationals who entered the country. The number of foreign residents with alien registration permits also increased by about 35 per cent during the decade to 1.22 million and nearly 80 per cent of these were Asian.[80]

The settlement and integration of foreigners into Japanese society has been progressing continuously for many years and although the relative numbers of foreigners in Japan remain small in comparison with those for Western countries, to have foreign neighbours in Japan is no longer considered exceptional.

In 1992, 4966 Japanese-foreign couples were married and in more than 95 per cent of these marriages a Japanese man married a Filipino woman. According to the Japanese Embassy in Manila, the number of international marriages to Filipino partners has increased rapidly since 1988 and reached a total of 21 000 in 1992. Some possible reasons behind this increase are that Japanese men from rural areas, who have encountered difficulty finding brides at home owing to the severity of rural depopulation, have turned to other Asian countries in their search, with the support of the rural municipalities; also, in urban areas, less well educated Japanese men who are not employed by the more prestigious companies report difficulties in finding Japanese brides. In contrast, more Filipino women, affected by the constraints on the entertainment business in Japan, appear to have been willing to marry Japanese men in order to obtain working visas. Although the majority of the Filipino women have managed to adjust to the Japanese way of life, some suffer from stress and depression caused mainly by language and cultural barriers. In consequence, eight municipalities in the Mogami area of Yamagata prefecture, where about 110 foreign wives are registered, have started to provide services, such as arranging for them to take Japanese lessons, helping them renew their visas perhaps as part of the process of naturalization, and locating appropriate churches to attend.[81]

The number of foreign children enrolling in local Japanese elementary and junior high schools has also been increasing, and most of them encounter serious language problems.[82] In the case of Aichi, Kanagawa, and Shizuoka prefectures, where approximately 6000 Japanese-South American descendants work in the car manufacturing industry, there is an increasing number of Japanese-Brazilian children who speak only Portuguese. To help them overcome this language barrier, the local Educational Councils have been trying to provide teachers who understand Portuguese and also to make Japanese conversation textbooks for these children.[83] The Tokyo metropolitan government has made a similar effort, providing the municipal boards of education in the Tokyo area with a list of volunteer interpreters, as well as copies of daily-conversation textbooks in six languages accompanied by Japanese translations.[84]

Notwithstanding these efforts by local government to facilitate the integration of foreigners into Japanese society, racial discrimination against foreigners, particularly Asians, continues. At the root of this discrimination is the fact that Japan is a highly self-contained society and in response to the heavy influx of illegal migrant workers, social conflict between native Japanese and the illegal migrants is developing.

Foreigners from Asian countries often experience difficulties in finding accommodation in Japan because many landlords refuse to rent to any non-Japanese including Korean and Chinese permanent residents. Many Japanese people view foreigners as troublemakers, and in particular try these days not to mix with foreigners for fear, it is alleged, of contracting AIDS. In 1992, 493 cases of AIDS (that is, people who had been tested HIV-positive) were reported in Japan – almost double the number recorded in 1991,[85] and among the 69 people who had contracted full-blown AIDS, 40 were South-East Asian women working in the sex industry.[86] In September 1991, a demonstration against foreign prostitutes was held in the Shinjuku area of Tokyo where a large number of brothels are located, the first organized display of feeling against the presence of illegal migrant workers.[87]

With the economic recession set to continue into 1994, there is some concern that illegal migrants could be taking jobs away from indigenous workers. One incident suggests the beginnings of a racist movement against illegal migrant workers. In April 1993, a neo-Nazi group called 'The League of State Socialists' began a campaign aimed specifically at excluding illegal migrant workers from Japan, especially Iranians, most of whom tend to gather in Yoyogi park. This group has erected large posters displaying swastikas with captions such as 'drive out illegal migrant workers' and 'recover Yoyogi park for the Japanese'. In addition to this poster campaign, a number of vehicles with loudspeakers have been cruising the streets of Tokyo calling for foreigners without visas to be expelled. So far, the Japanese government has taken no action against this neo-Nazi group.[88]

Recently, conflicts between different ethnic groups of illegal migrant workers have also emerged. Although these conflicts are currently mainly of an individual nature, disputes between Thai female illegal workers and their Filipino counterparts, and between Korean-Japanese descendants and Korean illegal migrant workers have been reported.[89] Many Korean illegal migrant workers residing in Kamagasaki, a slum area of Osaka, are also reportedly moving out of the network of Korean permanent residents for whom many of them work. These Korean-Japanese employers, who are accustomed to the Japanese-style priority of work over family relationships, often come into conflict with their employees from Korea who consider kinship to be more important than work. In the case of illegal migrant workers in Japan, no major conflicts among different ethnic groups have so far been reported since all groups remain afraid of exposure. The very existence of ethnic groups, however, poses serious challenges to the persistent Japanese concept of a homogeneous society.

CONCLUSIONS

Japan is the only country among the world's developed nations to have achieved high levels of economic growth without relying on foreign workers.[90] However, the recent massive expansion of the Japanese economy has generated strong demand for migrant workers.

This economic expansion can be explained, at least in part, in terms of the significant changes that have been taking place in the country's employment structure since the mid-1970s. The number of jobs that are necessary to support the mass domestic production system has increased in parallel with expansion of the economy. At the same time, the degree of labour specialization in various industries has correspondingly expanded. In particular, there has been a major growth in the advanced service industry, the successful operation of which requires an infrastructure of low-wage employment which is often accompanied by undesirable working conditions. It is this type of employment that is unattractive to indigenous workers and in their place, a large number of illegal migrant workers have filled the vacancies despite official policies designed to exclude them and encourage selected foreigners to work legally in the country.

However much the government and employers may wish to regard migrant workers as a flexible and low-cost response to labour shortages any short-term benefits are beginning to be outweighed by long-term social costs. The impact of the ongoing recession on the labour market has begun to diminish opportunities for illegal migrants to find jobs easily. Under the current economic climate, migrant workers are considered as surplus to requirements and have been the first to be fired as companies carry out massive cost-cutting programmes. Corporate giants such as Toyota Motor Corporation and Matsushita Electric Industrial Company have begun drastically cutting production to eliminate excessive, costly inventories. These strategies of course mean that orders to smaller subcontractors must also be cut back.[91] In order to reduce production, many subcontractors have been decreasing the number of working hours as well as reducing their workforces. As a result more illegal migrant workers are becoming jobless. It has been commonplace for large numbers of illegal migrant workers to be fired without notice and to be owed wages which may be in excess of a million yen (about £6000), representing several months' pay for an individual.[92] Strong fears exist in Japanese society that, in situations where illegal migrant workers lose their jobs and do not have sufficient money to return to their homeland or even to eat, they might take to crime or cause social problems.[93] Under current conditions, it seems likely that the Japanese government will soon have to shoulder significant social costs in respect of illegal migrant workers.

Owing to the presence of illegal migrant workers, it is true that Japanese workers are largely exempt from unpleasant, dirty and dangerous work. The illegality of the migrants means that there is little incentive to provide them with even basic human rights, such as social security benefits and protection under the law, and there is no fear of their building unsightly slums since they have to live silently in small groups, constantly in fear of being found out and deported. In a way, the Japanese people and the government have themselves been reaping the benefits of this illegality which might be considered to be an extreme case of institutional discrimination underpinned by common feelings among Japanese people of discrimination against other Asian people.

Japan is not the only country juggling with the problems of how to deal with its illegal migrant workers and more general issues relating to migration. Fears that a continued influx will give rise to social tensions and ethnic unrest, as well as political and economic problems, are commonly shared by other industrialized countries. In the case of Japan, however, discrimination against other Asian people adds to the tension. Further consideration of the complex dynamics of discrimination is necessary if the desire to make Japanese society more open to foreigners is to become reality. The government must act to build up an awareness that illegal migrant workers, as well as other minorities in Japan such as Korean and Chinese permanent residents, must be accorded human rights equal to those of the Japanese themselves. To grant these human rights will be only the first step towards solving the problems associated with illegal migrant workers in Japan.

NOTES AND REFERENCES

1. For analysis of the role of transnational enterprises with respect to North–South relations, see K.P. Sauvant and B. Mennis, 'Sociocultural Investments and the International Political Economy of North–South Relations: The Role of Transnational Enterprises', in K. Kumar (ed.), *Transnational enterprises: Their Impact on Third World Societies and Cultures*, (Colorado: Westview Press, 1980) pp. 275–92.
2. S. Sassen, *The Mobility of Labor and Capital – a Study in International Investment and Labor Flow* (Cambridge: CUP, 1988), p. 97.
3. Ibid., p. 98. See also Sassen 'Immigration Policy Toward Mexico in a Global Economy', *Journal of International Affairs*, (Winter 1990), p. 369.
4. Henceforth in this paper these 'illegal foreign migrant workers' in Japan will be referred to as 'illegal migrant workers'.
5. T. Iyotani, 'Sakerarenai Kadai' [Unavoidable Task] in T. Iyotani and T. Kajita, (eds), *Global Approach, Gaikokujin Rodosharon* [Global Approach, the Theories of Foreign Migrant Workers] (Tokyo: Kobundo, 1992), p. 102.
6. Ibid., p. 103.
7. Ibid.
8. Henceforth in this paper the 'Immigration-Control and Refugee-Recognition Act' will be referred to as the 'Immigration Law'.
9. The total number of foreigners who were working within the scope of the immigration requirements was 49 384 at the end of 1989 – this figure represents a 250 per cent increase on the figure for 1984. Immigration

Bureau, Ministry of Justice in Japan (eds), *Employment of Foreign Nationals: Questions and Answers*, (Tokyo: Nihon Kajo Shuppan, 1988).

10. The entry into and conditions of residence in Japan of all foreigners, including those who are to be employed, are regulated by the Immigration Law. For further explanation, see Immigration Bureau, *Employment of Foreign Nationals* (1988), p. 3.
11. Ibid., pp. 58–61, and in particular, the section entitled 'On Violations'.
12. Y. Sellek and M. Weiner, 'Migrant Workers: The Japanese Case in International Perspective' in G. Hook and M. Weiner (eds), *The Internationalization of Japan*, (London: Routledge, 1992), p. 224.
13. This was due to the relaxation in Taiwan of restrictions on overseas travel.
14. M. Matsuda, 'Kokusaibaikaishun no naka no Joseitachi' [Women in the International Prostitution Scheme] in Ajiajin Rodoshamondai Kondankai (ed.), *Okasareru Jinken, Gaikokujin Rodosha* [Exploited Human Rights, Illegal Migrant Workers] (Tokyo: Daisan Shokan, 1992), p. 88.
15. The total number of sending countries rose from 48 in 1990 to 75 in 1991 and included 21 African countries. 'Record 32 908 Workers Deported in '91'. *The Daily Yomiuri*, 9 June 1992.
16. H. Komai, 'Fuhoshurosha to iu Rakuin' [The Tag of Illegal Migrant Worker], *Japanese Economist*, vol. 67, no. 36, (1989), p. 75.
17. The intention was to promote mutual tourism with the oil-rich country, as well as to help maintain a reliable petroleum source. Y. Itoh, 'Visa Requirement for Iranians to be Temporarily Reinstated' in *The Nikkei Weekly*, 28 March 1992, vol. 30, no. 1510, (Tokyo: Nihon Keizai Shinbun, 1992).
18. Domestic economic conditions in Iran deteriorated because of the Iran–Iraq War, the unstable political situation following the Islamic Revolution, and the earthquake in November 1991. T. Mitsui, 'Gaikokujin Rodosha o Haijosuru Nippon' [The Japan that Excludes Foreign Migrant Workers] in Asiajin Rodosha Mondai Kondankai (ed.), *Okasareru Jinken, Gaikokujin Rodosha* [Exploited Human Rights, Illegal Migrant Workers] (Tokyo: Daisan Shokan, 1992), p. 23.
19. K. Tezuka, *Zoku Gaikokujin Rodosha* [Foreign Migrant Workers, continued] (Tokyo: Nihon Keizai Shinbunsha, 1991), p. 33.
20. The Immigration Bureau arrived at this figure by counting entries and departures of foreign nationals at airports and seaports. By nationality, Thais topped the list at 53 219, followed by 37 491 South Koreans, 34 529 Malaysians and 34 295 Filipinos. Because the government suspended its visa-exempt agreement with Iran, the number of Iranians who were in Japan illegally decreased from 40 001 in May, 1992 to 32 994 in November 1992. There were also 29 091 Chinese who had overstayed their visas, as well as 8161 Bangladeshis, 8056 Pakistanis and 7283 Taiwanese. The number of Peruvians more than doubled to 6241 from six months earlier.
21. 'Illegal Foreigners Total Nearing 300 000 Mark', *The Japan Times*, 25 February 1993.
22. 'Immigration Staff Fight Losing Battle', *The Japan Times*, 5 January 1993. In 1991, the largest number of foreigners refused entry was Iranian (7540), followed by Thai (6738) and Malaysian (4875).
23. Since the revision of the Immigration Law the restriction regarding the number of hours pre-college students are permitted to work has been tightened. Prior to the revision they could work up to 20 hours per week.

24. H. Komai, 'Meisosuru Nihongo Gakko Taisaku' [Policy Against the Problem of Japanese Language Schools], *Japanese Economist*, vol. 67, no. 39, (1989) p. 58.
25. Ibid., p. 58.
26. H. Tanaka, *Zainichi Gaikokujin* [Foreign Residents in Japan], Iwanami Shinsho: 209, (Tokyo: Iwanami Shoten, 1991), p. 75.
27. H. Komai, 'Meisosuru Nihongo Gakko Taisaku' p. 58.
28. H. Komai, 'Zodaisuru Chugoku no Romuyushutsu Atsuryoku' [Increasing Pressure to Export the Labour], *Japanese Economist*, vol. 68, no. 1, (1990) p. 113.
29. H. Tanaka, *Zainichi Gaikokujin*, pp. 166–74.
30. Homusho Nyukoku Kanri Kyoku [Immigration Control Bureau, Ministry of Justice] (eds), *Showa 63 Nen Kamihanki ni okeru Kyohisha oyobi Nyukanho Ihan Jiken no Gaikyo ni Tsuite* [On Cases of Refusal of Landing Permission and Violations of Immigration Law in the First Half of 1988] (Tokyo: Okurasho Insatsukyoku, 1988).
31. J. Goto, 'Gaikokujin Rodosha Ukeire no Keizaigakuteki Eikyo' [Economic Effects of the Acceptance of Foreign Migrant Workers] in *Kikan Rodoho*, no. 164, (Tokyo: Sogo Rodo Kenkyujo, 1992), p. 33.
32. As note 30.
33. K. Tezuka, *Zoku Gaikokujin Rodosha* [Foreign Migrant Workers, continued] (Tokyo: Nihon Keizai Shinbunsha, 1991), p. 70.
34. M.J. Piore, *Birds of Passage* (Cambridge: CUP, 1979), p. 4.
35. Ibid., p. 1.
36. These factors were suggested by the Board of the Economic Planning Committee, Ministry of International Trade and Industry in Japan (eds), *Gaikokujin Rodosha to Keizaishakai no Shinro* [Migrant Workers and their Impact on the Economy], (Tokyo: Okurasho Insatsukyoku, 1989), pp. 41–50.
37. World Bank (eds), *World Development Report* (Washington DC, 1990).
38. In the 1960s approximately 10 million labourers were supplied from rural areas in Japan. See K. Morita and T. Iyotani, 'Japan and the Problem of Foreign Workers' (Discussion Paper), Faculty of Economics, University of Tokyo (1990).
39. The number of seasonal workers was approximately 600 000 at the beginning of the 1970s, but owing to the increase in job opportunities generated by the economic development of rural areas and the concurrent depopulation of these areas, this number decreased to 200 000 in 1989. See Goto, 'Gaikokujin Rodosha Ukeire no Keizaigakuteki Eikyo', p. 35.
40. Y. Sellek and M. Weiner, 'Migrant Workers', p. 220.
41. T. Iyotani, 'Sakerarenai Kadai' [Unavoidable Task] in Iyotani and Kajita (eds), *Global Approach*, pp. 118–19; Sellek and Weiner, 'Migrant Workers', p. 221.
42. Goto, 'Gaikokujin Rodosha Ukeire no Keizaigakuteki Eikyo', p. 35.
43. K. Tezuka, *Zoku Gaikokujin Rodosha*, p. 43.
44. Ministry of Labour (ed.), Rodo Hakusho, Heisei Yonenban [White Paper on Labour in 1992] (Tokyo: Nihon Rodo Kenkyu Kiko, 1992), p. 26.
45. 'Financial Therapy for a Mid-life Crisis', The *Financial Times*, 5 February 1993.
46. As note 44.

47. T. Hachiya, *Soredemo Gaikokujin Rodosha wa Yattekuru* [Even so, the Foreign Migrant Workers are Still Coming to Japan] (Tokyo: Nikkan Kogyo Shinbunsha, 1991), pp. 65–7.
48. Ibid., p. 70.
49. Public Employment Security Office, Ministry of Labour (eds), *Gaikokujin Rodosha Mondai no Doko to Shiten* [Trends and Viewpoints Relating to the Issue of Migrant Workers in Japan] (Tokyo: Romu Gyosei Kenkyujo, 1991) p. 35. In terms of supply and demand for labour, the situation in developing countries is in marked contrast to that in Japan. For example, in the decade from 1990 to 2000, Bangladesh and Pakistan, with estimated annual labour-force growth rates of 3.6 per cent and 3.7 per cent respectively, will need to create between 10 and 12 million new jobs each year to retain current levels of employment. Without a wave of foreign investment which would have to be larger than that experienced by ASEAN in the 1980s, these countries face either rapidly growing unemployment or the need to place far more workers overseas. See L.D. Rosario and G. Fairclough, 'Toilers of the East', in *Far Eastern Economic Review*, 2 April 1992, p. 20.
50. M. Asano, 'Kensetsu Rasshu no Seika, Ote ga Hitorijime' [Result of the Boom in the Construction Industry – Major Benefits go to Large Companies], *Rodo Undo*, no. 291 (1990), p. 128.
51. Kensetsugyo Gaikokujin Mondai Kenkyukai (eds), *Kensetsugyo ni okeru Gaikokujin Rodosha Mondai to Gaikokujin Kenshusei no Ukeire* [The Problem of Foreign Migrant Workers and the Acceptance of Foreign Trainees in the Construction Industry] (Tokyo: Taiseishuppansha, 1991), p. 11.
52. Sellek and Weiner, 'Migrant Workers', p. 216.
53. Ibid., p. 217.
54. M. Matsuda, 'Baikaisareru Ajia no Onnatachi' [Asian Women for Sale] in *Hogaku Semina Zokan, Gaikokujin Rodosha to Jinken* [Hogaku Seminar Special Issue – Migrant Workers and their Human Rights], vol. 42, (1988) pp. 36–8.
55. Hachiya, *Soredemo Gaikokujin Rodosha wa Yattekuru*, p. 77.
56. Japan Immigration Association, *A Guide to Entry, Residence and Registration Procedures in Japan for Foreign Nationals* (Tokyo: Nihon Kajoshuppan, 1990), pp. 26–37.
57. Y. Koido, *Gaikokujin Rodosha – Seisaku to kadai* [Foreign Migrant Workers – Policies and Tasks] (Tokyo: Zaimu Keiri Kyokai, 1990), p. 41.
58. Ibid., p. 42.
59. Ibid., pp. 49–50.
60. Ibid., p. 48.
61. Japan Immigration Association, *A Guide to Entry*, p. 33.
62. Ordinance no. 16 of the Ministry of Justice (1989).
63. In general there are three types of organization that accept foreign trainees: organizations to which the government pays the entire expenses of the trainees, organizations for which the government and private companies share these expenses, and private organizations. It is these last that are suspected of making trainees work for very small remuneration. Hachiya, *Soredemo Gaikokujin Rodosha wa Yattekuru*, p. 115.
64. Hachiya, *Soredemo Gaikokuhjin Rodosha wa Yattekuru*, p. 112; and Development of Employment Section, Ministry of Labour, 'Aratana

Gaikokujin Kenshu Seido', in *Kikan Rodoho*, no. 164 (Tokyo: Sogo Rodo Kenkyujo, 1992), p. 87.

65. Ibid., pp. 87–94.
66. Hachiya, 'Gaikokujin Rodosha Mondai Ukeire to Seifu, Keizaikai no Tachiba' [The Acceptance of Foreign Migrant Workers and the Stances Taken by Government and Industry in Japan] in *Kikan Rodoho*, no. 164, (Tokyo: Sogo Rodo Kenkyujo, 1992), p. 55.
67. The average monthly allowance for trainees in the construction industry is approximately 60 000 yen (about £350).
68. H. Komai, 'Teichingin Rodosha toshite no Gijutsu Kenshusei' [Trainees as Labour with Low Wages], *Japanese Economist*, vol. 67, no. 40, 1989, p. 66.
69. Under the previous Immigration Law, the third generation of South American-Japanese descendants was granted a status of residence known as 'designated activities' in order to qualify for work.
70. Sangyo Koyo Antei Centre (eds), *Nikkeijin Koyo no Kiso Chishiki* [Basic Knowledge of Employing South American-Japanese Descendants] (Tokyo: Rodo Shinbunsha, 1991), p. 10.
71. On average, South American-Japanese descendants earn 1300 yen per hour, whereas the hourly wages of illegal migrant workers are much lower. For example, it is estimated that illegal workers from Pakistan and Bangladesh earn 900 yen per hour, followed by those from Sri Lanka, Nepal, and Ghana at 800 and Iranians at 600–700 yen per hour. See A. Hinago, *Kei, Nen, Kokuseki Fumon – Gaikokujin Rokosha ga Hyakumannin ni Naru Hi* [No Restriction on Age, Nationality, and Experience – the Day When the Number of Illegal Migrant Workers in Japan Reaches One Million] (Tokyo: Diamond Sha, 1992) p. 114.
72. M. Yoshimen, 'Nikkeijin no Wagakuni ni Okeru Shuro no Genjo to Taisaku' [The Employment Situation and Measures Relating to South American-Japanese Descendants] in *Kikan Rodoho*, no. 164, (Tokyo: Sogo Rodo Kenkyujo, 1992), p. 114.
73. Hachiya, *Soredemo Gaikokujin Rodosha wa Yattekuru*, pp. 156–61.
74. Ibid, pp. 144–55.
75. These examples of violations of human rights are taken from Y. Matsui, 'Jinken Shingai ni Naku' [Tears over the Violation of Human Rights] in A. Utsumi and Y. Matsui (eds), *Ajia Kara Kita Dekasegi Rodoshatachi* [Migrant Workers from Asia] (Tokyo: Akashi Shoten, 1988).
76. Illegal migrant workers are subject to various Japanese labour-related laws, such as the Labour Standards Law, the Minimum Wage Law, the Labour Safety and Sanitation Law, the Labour Union Law, the Employment Security Law, the Employment Insurance law, and the Workmen's Accident Compensation Insurance Law. Article 3 of the Labour Standards Law makes it illegal to discriminate against any worker on the basis of nationality in respect of wages, working hours and, in many cases, working conditions. Y. Oshimo, *Gaikokujin Koyo* [Employment of Foreigners], (Tokyo: Nikkeiren Kohobu, 1991), p. 44.
77. 'Newcomers Live with Fear, Stress', *The Japan Times*, 22 January 1993.
78. T. Takayama, 'Gaikokujin o Shimedasu Iryo, Fukushi' [Medical and Welfare Systems which Exclude Foreigners] in Ajiajin Rodoshamondai Kondankai

(ed.) *Okasareru Jinken, Gaikokujin Rodosha* [Exploited Human Rights, Illegal Migrant Workers] (Tokyo: Daisan Shokan, 1992), p. 132.

79. M. Selby, 'Human Rights and Undocumented Immigrant Workers in Japan', *Stanford Journal of International Law*, 1989 (Fall) vol. 26, part 1, pp. 325–69, gives a comprehensive explanation of the human rights of illegal migrant workers.
80. 'Asian Newcomers Integral to Society', *The Japan Times*, 7 January 1993.
81. 'More Wed Filipinos, Reaching New Record', *The Japan Times*, 30 January 1993.
82. Official statistics released in May 1992 indicate that 686 elementary school pupils in Tokyo required extra Japanese language lessons, compared with 416 in the previous year. In junior high schools, these figures were 396 and 278 respectively. 'Plan to Ease Language Barrier', *The Japan Times*, 19 January 1993.
83. H. Tanaka, *Zainichi Gaikokujin* [Foreign Residents in Japan], Iwanami Shinsho: 209, (Tokyo: Iwanami Shoten, 1991), p. 209.
84. 'Plan to Ease Language Barrier', *The Japan Times*, 19 January 1993.
85. 'AIDS is No Longer a "Foreigners" Disease', *The Japan Times*, 21 January 1993.
86. 'AIDS and HIV Cases Double', *The Japan Times*, 27 January 1993.
87. R. Itoh, '"Japayuki-san" Gensho Saiko' [Re-examination of Japayuki-san Phenomenon] in Ityotani and Kajita, (eds), *Global Approach*, p. 326.
88. T. McCarthy, 'Racists Cast a Shadow on the Rising Sun', *The Independent*, 12 April 1993.
89. Interview with Mr Ashura Mizuno, leader of a civic group called 'Asian Friends'; conducted in summer 1992.
90. In this case, 'foreign workers' do not include Korean and Chinese permanent residents in Japan.
91. It has been estimated that orders to small manufacturers have been reduced by more than 20 per cent on average and in addition subcontractors in the automobile industry have been required to cut the costs of parts by 50 per cent since December 1992. *Nikkei Weekly*, 21 March 1992, vol. 30, no. 1509, 'Slump puts Labour Shortage in Lull'.
92. Foreign Workers' Woes Rise in Protracted Slump', *The Japan Times*, 16 March 1993.
93. 'Slump puts Labour Shortage in Lull', *The Nikkei Weekly*, 21 March 1992.

9 Relocation in Vietnam and Outmigration: The Ideological and Economic Context

Linda Hitchcox

This chapter examines migration in and from Vietnam, exploring the ways in which the phenomenon is constructed by participants and authorities as political ideology, as economic necessity and as a social support system. The last 15 years provide the focus, but much of the relevance of the following discussion depends on the background: the chaotic wartime conditions when many millions of Vietnamese moved from their homes.

In the autumn of 1954, immediately prior to the partition of the country as agreed in the Geneva Accords, more than 850 000 people, mainly Catholics and small landowners, journeyed south, while 80 000 of those loyal to the new Communist government travelled in the opposite direction. In the following years until unification in 1975, it is estimated that in South Vietnam more than half the population of 20 million was displaced, usually because of the bombing or the strategic hamlet programme.[1] The uprooted tended to drift to the cities[2] especially Ho Chi Minh City (Saigon). In the north, similar processes were under way with many thousands evacuated to the countryside to escape the bombing, and a further million people resettled from the Red River delta to new economic zones on the borders of the Democratic Republic of Vietnam.[3] Since unification of the country in 1975, signalling the end of the war against the American-backed government of South Vietnam, there have again been large-scale internal and external movements of Vietnamese, mainly in response to the socialist political and economic policies issuing from central government.[4]

There are both legal and illegal forms of outmigration. With regard to the former, more than 300 000 workers have gone abroad temporarily in the 10 years since 1981, mainly to socialist countries organized through a Labour Co-operation Agreement signed in 1981 that helped to service debts outstanding to the Soviet Bloc.[5] Each year, they were joined by several

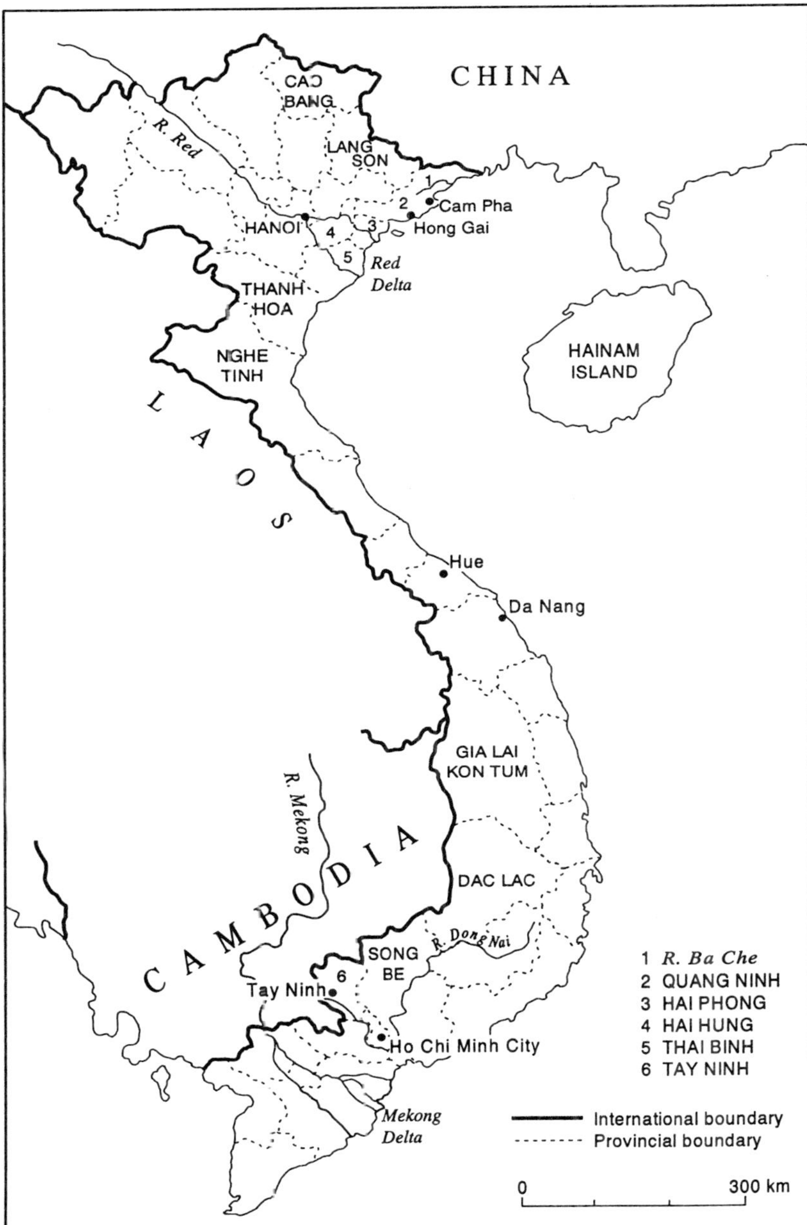

Map 9.1 Vietnam

thousand students who often stayed several years in the former Soviet Union, to gain training unavailable in Vietnam. More than 170 000 Vietnamese have obtained work or education in the USSR and in 1990 there were still 60 000 people working and living there. Until that year, a Moscow degree served as a status symbol and almost as a prerequisite to any successful career in the government, Party or professions.

Illegal outmigration of asylum seekers and refugees continued for many years after 1975 – not surprisingly, given the bitter, costly ideological struggle which was both a civil and an international war. More than a million people left Vietnam and found a sympathetic welcome in Western countries such as Australia, Canada and the USA, resettling there as refugees.

By the late 1980s, however, attitudes had changed. Doubts were growing within immigration departments over the justification of continuing to grant refugee status to Vietnamese who were still arriving in the reception centres in South-East Asia. Screening interviews were introduced in 1988 and the following year a repatriation programme was established, to allow those who failed to be granted refugee status to return to Vietnam, if they agreed. In a study conducted in the Hong Kong detention centres in 1990, just under 60 per cent of the sample were found to be not justified in claiming asylum on grounds of oppression or persecution.[6] At that time, the screening process was rejecting approximately 90 per cent of applicants. The percentage has since adjusted to present levels of approximately 70 per cent rejection in the region overall. Among those rejected, more than 8000 asylum seekers have returned to Vietnam. They form one of the groups of relocated people discussed below.

Internal migration is a common experience in Vietnam. Since 1975, several million people have officially or unofficially moved from one region to another, usually going to new economic zones. The context is explored here by first referring to the political economy of Vietnam since 1975. Next, attention is directed to relocation as both political policy and economic strategy, noting in particular the circumstances of Hmong people and others after the closure of the co-operatives in 1988. Lastly, for comparative purposes, the discussion turns to fieldwork data drawn from interviews with 50 asylum-seeker returnees from Hong Kong now living in the northern coastal province of Quang Ninh. Tentative conclusions are drawn concerning the likely connections between the occurrence of movement from one province to another and the preparedness to migrate abroad, proposing that it is instability and uncertainty about the future more than economic hardship that has swelled the numbers of people leaving Vietnam.

THE ECONOMIC AND POLITICAL BACKGROUND

The Vietnamese economy is typically described as being in a state of transition.[7] Since 1988 it has been moving from a highly centralized command economy to a more decentralized system in which the private sector has a vital part to play and government subsidies, food rations, and subsidiary means of production have been largely withdrawn.

This radical change need not necessarily be seen as a sudden turnaround, but rather a more realistic admission by the government of the presence of underlying economic forces. There are those who would argue[8] that the role of the state was always weakly applied to economic policy and the present transition is no more than a tacit recognition of the failure of the centre to control enterprise at the village or commune level. Throughout the hardships and constraints of the war years, the state was obliged to accept in practice what it denied in theory: some channelling away of rice and other resources as peasants sought better prices for their goods on the free market. From this perspective it might be said that in the last few years, the state has recognized the strength of an economic trade that first began to surge in the 1960s in the Democratic Republic of Vietnam. It was temporarily bridled to a Stalinist-style programme of a united victorious Vietnam in the late 1970s and began to emerge once more after 1981, with the admitted failure of the industrialization programme and collectivization of agriculture in the south of the country. This failure mainly stemmed from the inability of the state, deprived of Western and Chinese aid, to maintain tight control of the economy when poor harvests under collectivization were pushing up the price of rice on the free market, further undermining the co-operatives.

Viewing the so-called transition rather as an open recognition of already-established features of the economy has much to recommend it. It offers an explanation of first the apparent rapidity of the adjustment to petty trading enterprise in the last three years, observable nearly everywhere in Vietnam, and second, the continuity of existing power relations of the cliental type, within communes that have seen the total disappearance of the co-operative system. At the level of everyday experience, the average citizen recognized the economic contradiction faced by many government officials who found themselves unable to survive on a state salary and relying on their energy and wits for a livelihood contrived outside formal employment. For those thus engaged, legitimation of such outlets in the late 1980s was a rationalization of their situation rather than a revolution.

The formal policy shift came in the Sixth Party Congress held in December 1986, where central planning was openly criticized, subsidies withdrawn, the role of the co-operatives radically reduced and the borders opened up to foreign trade. The immediate impact of the withdrawal of government subsidies was on the reduction in state employment by 500 000 between 1988 and 1989. In 1991, some 50 per cent of state enterprises were reported to be effectively bankrupt, but a considerable sector was responding to market forces, had become more streamlined and was successfully operating at high profitability. Foreign trade expanded rapidly in 1989, largely through informal contacts, demonstrating once again weaknesses in the central authority of government and its reluctance to become decisively involved in making ideologically-loaded decisions concerning the managed development of a free market.

The collapse of trade contracts with Eastern Europe, alongside the gradual reduction of Soviet aid, also considerably weakened any possible room for manoeuvre by central government. Estimates of aid from the USSR in 1986 were US$1.8 billion, but by 1990 this assistance and that from all the CMEA (Council for Mutual Economic Assistance) countries had been severely curtailed. The urgent priority for Vietnam's leaders was not only to obtain development aid elsewhere, but also to establish alternative trade outlets to accommodate the 70 per cent of foreign trade that in 1988 lay with CMEA countries. By 1991, some small successes had been achieved with EC and ASEAN countries, despite the delay in resolving the Cambodia issue, then the main reason declared by the USA for the embargo on IMF and World Bank loans to Vietnam.

At the Seventh Party Congress held in June 1991, the indications were that the Vietnam government was attempting to remain in control of all the economic currents that had been unleashed in the previous three years without moving very far in terms of political change. The emphasis was on a controlled (albeit weakly implemented), rather than a managed economy. One small example was the introduction in 1991 of a 25 per cent tax on all goods moving from south to north – an attempt to raise revenue and control traffic, but with little account taken of the importance of this movement of goods in generating a stronger market economy.[9] With the economic changes formalized at the 6th Party Congress, there also developed some relaxation of the more Stalinist-style policies of control affecting social and public life.

As a factor affecting migration, one of the more important changes was the lifting of restrictions on employment and freedom of movement. Before January 1987, it was compulsory for all citizens to be employed; which meant, almost by definition, by the state. Those without work and of politically acceptable background were offered jobs in a region where there was employment available: the forests, the coalmines or other industries and the new economic zones or development areas. In 1991, there were still professionals who, 14 years earlier, had reluctantly agreed for the sake of their country to work hundreds of miles from home and family, and who urgently wished to return, but still had not obtained permission from their employers.

Movement for the purpose of short visits also used to be tightly controlled. The procedure for travelling or visiting any other place in Vietnam first involved going to the employer for permission. If s/he agreed, a paper was provided that had then to be confirmed by the local police and stamped. On arrival at the destination the paper had again to be stamped by local police and once more at the end of the visit. On return, the paper was shown to the employer who returned it to the local police for a final stamp. No business or trade was possible between provinces because of these restrictions and each province had its barrier and patrolling police. After January 1987 and the introduction of freedom of movement, it was no longer compulsory to have a job and tradespeople could conduct business, including transporting goods between provinces.

It is still the case that people tend to write letters to the authorities requesting the right to move somewhere else although this is no longer necessary; but they do it in order to play safe, in case the rules are changed while they are in transit. People do have to remain registered in their home district for welfare and health benefits, but since the decline of the co-operatives such services have been so severely reduced that this right does not act as an important constraint on movement.

The considerable changes in the economy mentioned earlier, while liberating for some, have faced others, particularly state employees, with instability and insecurity. In the early years of change they were accompanied by shrinking opportunities of obtaining state-sanctioned employment and training in Eastern Europe and the former Soviet Union. For some, the unofficial labour migration alternative has been to move rather further than a neighbouring province and follow the route of refugees to Hong Kong and elsewhere.

Overall, the broader context is one of fluidity and opportunity, of a lessening of state control in some areas of public life (at best unevenly and weakly applied), of instability with growing unemployment particularly in the state sector, and accompanying disillusionment at a time of the general weakening and frequent collapse of the traditional care services.

This is the background to internal migration, an option that is often demanded and planned by the state which then, more often than not, leaves people to their own devices in making relocation successful. The habit of movement is thus reinforced as politically acceptable behaviour in Vietnamese socialist society, but alongside self-reliance.

INTERNAL MOVEMENT IN VIETNAM

In 1975, the newly-unified government faced extremely serious problems arising primarily from the dislocation of the country's population. Before the end of the war in Vietnam in 1975, it is estimated that more than 10 million people had been forced to abandon their homes,[10] because of bombing, residence in targeted combat zones, or because of loyalties to the governments of the South or North that could divide communities and even families. The major priority of the government, once the war ended, was to organize and increase food supplies and accordingly there was a rapid expansion of new economic zones aimed at distributing land more evenly among the population from the overcrowded delta regions of Mekong and Tonking.

The defeated had little choice about their relocation and a common report from refugees/asylum-seekers is of being sent to cut out a living in a border area with almost no support from the government. For the victors, the element of choice was not greatly more extensive. More than 230 000 veterans were transferred from the army to work for economic construction. District officials in a mountain district of Quang Ninh reminisced about the appeal that had been made to their patriotism after they had returned from the war, how proud they had felt to offer the sacrifice of leaving their home regions to establish a New Economic Zone, and the loneliness and hardship of the succeeding years.

The first redistribution phase of the government population programme began in 1975 followed by the second phase of the same programme in 1981 and still in progress, although on a much reduced scale. The most important feature of the present situation is the movement to the Central Highlands of Vietnam, from almost every other region, with the emphasis on control of urban growth. According to the statistical office in Hanoi, the size of the urban population at 19.8 per cent (1989) of 64 million people is still below the figure of 20 per cent registered in 1976. In the first four years after the war ended, more than two million people were officially relocated to another region (2 207 811), including 624 000 people from the south-east (Ho Chi Minh City among other places) of which 332 751 were relocated within the region (for example Song Be), and 598 000 out of the Red River Delta mainly to the Central Highlands.[11]

Over the following 10-year period, from 1979 to 1989, the predominant characteristic was of movement from the north (mainly the Red River Delta) to the south and usually to the border regions. Taking into account natural increase there was an outmigration of 185 877 people from the Northern Highlands; 681 780 from the Red River Delta; 841 589 from the Northern Central region; 469 090 from the Central Coastal region, and

982 943 from the Mekong Delta. Only the south-east (Ho Chi Minh City, Dong Ngai, Song Be and Tay Ninh) and Central Highlands experienced net inflows of 283 478 and 233 029 respectively.[12]

Altogether, since 1976, the government has assisted more than four million people to move from one province or district to another. Noting the general direction of the movement, we may surmise that the government saw the advantages of mixing a population that had previously been divided more or less between north and south along ideological lines. The uncertain loyalties of some of the minorities were also more easily fixed with the arrival and settlement of northern lowland formers who had lived under communism since the 1950s.

The political motivations may be clear, but it is also important to remember the driving economic necessity of feeding the people in the period after 1975 at a time when Vietnam was largely isolated from the international community mainly as a result of the Vietnamese overthrow of the Khmer Rouge regime in Cambodia in 1978. In the Central Highlands, a region that had been notorious for FULRO (the united front for the liberation of oppressed races) activity, there are large areas of cultivatable land, some given over to coffee and rubber plantations where there is a good opportunity to make a worthwhile living. Relocation and voluntary resettlement has therefore continued intensively to this region, fuelled by the overcrowding and shortages of land on the low-lying wet rice plains comprising only 20 per cent of the total land area of Vietnam, but where 80 per cent of the population is concentrated. In Thai Binh, arable land is less than one square metre per person. (The UK has, by contrast, 4.3 square metres per person.) The development equation has usually been seen in the simple *quid pro quo* terms of 1.9 million surplus labourers in the Red River Delta and an estimated 753 000 labourers required in the Central Highlands.

In areas of some farming potential, the main problem lies with providing adequate support for at least the first six months after resettlement. Each officially relocated family receives indirectly some three million dong, about US$450, an insufficient sum when a buffalo can cost approximately 500 000 dong and a simple block house more than one million.[13] Sufficient funds are not available to support the 50 projects that the government has planned to help such areas, particularly in the Central Highlands where there is plenty of land but few facilities. Each government project is planned to last five years at a cost ranging from three to 15 billion dong[14] reaching out to more than one million people. The average yearly planned movement is 250 000 to 300 000 people costing the government some 30 billion dong overall.

The government's original intention was to move between nine and ten million people, but problems of poor planning, inadequate services, unhealthy malarial regions and poor soil sustainability rapidly began to emerge among those already settled. The Department for the Reallocation of Population and Manpower's *own* surveys on New Economic Zones scattered throughout the country revealed that in 1989, after nine years of implementation, the circumstances of many of the settlers were bleak. Five years after resettlement, only 27 per cent of the total of migrant households reported to government officials that they had found a better life in all respects than before, 48 per cent declared their living was the same, and 25 per cent admitted finding conditions very difficult. Fifteen per cent of households in the chosen sample had returned to their place of origin.[15]

Additionally, partly because of the problems of low productivity and poor technology, the impact on the environment has been very negative, with an estimated five hectares of land in highland districts becoming subject to deforestation and erosion for every one hectare of land actually in the process of cultivation. The government admits that one of the main problems has been the attempt made by settlers, unskilled in any kind of agriculture, to grow cereals on unsuitable land, as part of the national drive to achieve local food self-sufficiency, at a time when there were prohibitions on freely exchanged production.

To sum up, the relocation programmes have proved unable to satisfy all the ideological and economic objectives for which they were intended to be the encompassing panacea. National policy was essentially concerned with security of borders, building and strengthening the idea and the reality of a united people in a unified Vietnam, of supporting the labour force and promoting the national campaign of local food self-sufficiency in all regions. It was also expected 'to be identical with the objectives of the regional development policy' through its programme of 'rationally redistributing productive forces'. Regional objectives were conceived as forming one integral part of the national migration programme, whereby some 12 million hectares of uncultivated land were to be developed at the same time as the ecological environment was to be conserved. The broad and sometimes contradictory character of these aims invited a failure that was inevitable in the pre-1987 conditions of local self-sufficiency, restrictions on movement and exchange of goods. Pioneering does not happily coexist within a state-controlled system.

SPONTANEOUS MOVEMENT

Freedom of movement was introduced in 1987 and since then, there has been a rapid increase in the numbers of people spontaneously moving from place to place, making any forward planning on the part of government (such as it expects to implement) increasingly difficult. In making a move, people now tend to rely more often on clan and kinship networks since the government has reduced assistance, and are more likely to be motivated by unstable conditions at the point of origin than by impoverishment. For example, in the typhoon-ridden coast of Nghe Tinh, the person who considers moving is the woman who is accumulating a little capital from rice wine production and pork, not her landless or incapacitated neighbour, for whom migration is not an option. From such poor provinces of the northern central region there is a tendency for people to seek work in industrialized areas such as the coalmining industry in Quang Ninh, as discussed in a later section.

The trend of frequent movement appears to be escalating. In the seven-year period from 1980 to 1987, approximately 100 000 people (20 000 households) moved of their own accord to the south. In the following three years to mid-1991, more than 700 000 such people were on the move. (This increase is echoed in the outmigration figures: 75 962 asylum seekers arrived in Hong Kong in the period 1987–91, compared with 5398 in the previous three years.)

Particularly in the northern border provinces, improved chances of profitable trade with China allow people to accumulate a little money, take decisions into their own hands and try their luck elsewhere. They do not move to employment but to what they perceive as opportunity. In a recent six-monthly period, more than 40 000 people left the provinces of Cao Bang and Lang Son on the Chinese border where local agricultural productivity is declining: in Cao Bang alone, nine out of the 11 districts have severe problems of food shortage.

MIGRANT-GENERATING REGIONS

Cao Bang province is an example of outmigration where falling soil fertility is a prime factor, but the more immediate catalyst is not overcrowding, since many thousands fled during the border clashes with China in 1979. Instead, the current impetus is more likely to have been the disappearance of the co-operative system in 1988 (it is relatively rare for co-operatives to cease functioning altogether), and the destabilizing consequences for the communities it served.

First of all, health, welfare and education services, deprived of both state and local support, have become so weakened as to be almost inoperable in some communes. Trained personnel and local officials are often the first to leave, deepening the predicament of those who remain. The second consequence is related to the handing back and dividing up of the land according to tenure rights existing at the time of French colonial rule, before the establishment of the co-operatives. There were inevitable anomalies in deciding ownership. Over the years from 1954 to 1988, more people had been included in the co-operative than there was land to give them, but they had been able to make a living by the practice of sharing all the resources of the co-operative. After redistribution, several families in each commune found themselves deprived of a livelihood and with no land to support themselves. Others were left with very little land, of poor quality, which was difficult to improve now that fertilizers were no longer available through the co-operative, and when irrigation systems, where they existed, were functioning less efficiently. More than two-thirds of the households in Cao Thuong (Hmong village) were in this situation.

A few farmers in each commune benefited, finding that their land had been considerably improved under the co-operative system and they were now able to buy up additional land from others less well placed, whose living was no longer viable, or plots which became available when schools and kindergartens were dismantled. The predominantly Nung commune at Ngoc Dong lost two of its four schools and a kindergarten with the closure of the co-operative, releasing some four hectares of land.

Those families with little land or none at all had no choice but to seek a living elsewhere; usually if they were already impoverished by moving a short distance further into the mountains or, if they had some money, by joining the steady exodus of people, which by 1991 was more than 30 000 from Thong Nong District alone, to provinces in the South such as Gia Lai Kon Tum and Dac Lac.

Community relations tended to weaken with disagreements over land ownership, thus adding to the general feeling of instability, throwing people back on family ties. It created a climate where people were inclined to take matters into their own hands, including migration among their strategies. In what was probably a temporary over-reaction to the close-down, neighbourly collaboration dramatically reduced, particularly among Nung and Tay. In the obedient response to government directives to 'do things for themselves now', people found that it was extremely difficult to arrange meetings or tasks requiring co-operative effort.

As it is generally the best-placed and most skilled who move first, the impoverished members of the community tend to be left at an even greater disadvantage. More land may become available, but poverty is often closely linked to lack of labour power in the family which means that more land cannot be exploited, even if it is there to be used. Only greater productivity, through better technology, can solve the problem.

For Hmong inhabitants living in Cao Thuong village (2000+), the idea of moving was a very radical and also expensive step, but food shortages have been driving them to consider the idea of breaking familiar and traditional ties. Forty households have already made the long journey south, one entire family succumbing to malaria only three weeks after arrival. One member of this household stayed behind to await the successful establishment of the new base, but now remains in the village, bewildered as to what to do next.

Nearly all of the 30 families interviewed in Cao Thuong could claim close kin relations within the village. One Hmong woman typically reported that she had relatives in eight out of the 80 other households, all of whom would consider moving if she did, a decision that might affect approximately 90 people. Traditionally, there is a preference for clan groupings in long-term migration.[16] A Hmong household usually consists of an extended family as in one case, where the parents, three sons, three daughters-in-law and their children were all living together in one house. It is rare that a member of the family is able to explore the new situation before the departure of the entire family or clan, unlike Tay and Nung. However, even if little news can be obtained, the presence of known relatives in another region acts as a link when the idea of moving is discussed, and may be the sole reason for considering one place rather than another, despite the fact that little other firm information may be available.

THE CENTRAL ZONE OF NGHE TINH AND THANH HOA

The central, coastal zone of Nghe Tinh and Thanh Hoa is also an area of net outmigration, an impoverished region, often hit by typhoon and floods, and with a dry inland plain, capable of sustaining only one crop of rice in most places. In Nghe Tinh, one of the largest provinces, with a population of some four million, the intention of local government is to promote a more even distribution of population by moving people from the most overcrowded, under-resourced communes lying in the typhoon zone along the coast, to zones of new development, usually forest districts. The ethnic minority peoples already occupying such areas are encouraged to abandon their shifting cultivation lifestyle and settle within the villages established by lowlanders. However, many new settlers do not have access to the range of resources that are necessary to support them in areas of thin soils and rising population. In 1990, 11531 people were formally recorded as moving from one district to another in Nghe Tinh, usually having already moved once and in search of a subsistence livelihood. In addition, in this one year, approximately 3000 people left the province altogether, numbered among more than 840 000 migrants who officially left this central region in the previous nine years, a significant number of whom moved to Quang Ninh.

MIGRATION IN AND FROM THE PROVINCE OF QUANG NINH

Quang Ninh is a province that has attracted large numbers of migrants and has been the exit point for considerable outmigration in the last few years – at least 18 000 people, all of whom have gone to Hong Kong in the hope of finding resettlement abroad. Quang Ninh lies on the northern coast of Vietnam adjacent to the Chinese border and is the centre of the largest coalmining industry in Vietnam. Road, rail and sea communications to and within the province are relatively good and a number of subsidiary and service industries are established around the port of Hong Gai.

Population movement is a common occurrence in Quang Ninh, with apparently significant connections between internal migration and outmigration to Hong Kong. There has developed an outlook among inhabitants that it is understandable and right that an individual should move in search of better prospects, an outlook that is implicitly endorsed by the government.

From 1980 to 1987, more than 54 000 people, mainly men, came to Quang Ninh recruited through government campaigns to work in the coalmines. Twenty-five thousand in this group came from other provinces, most commonly from the impoverished districts in Hai Hung and Nghe Tinh, usually in response to government recruitment schemes. More than 4000 wives and relations followed them at a later date when workers had established themselves in the mining communities and accommodation was available. In the town of Cam Pha alone, 5000 coalworkers arrived every year until 1987, the year when government economic planning dramatically changed direction. As part of the transition to a non-subsidized system, government recruitment schemes stopped and apprenticeships were no longer available for the children of mining families. At the same time, opportunities for employment in Eastern Europe vanished, together with contracts placed with Vietnamese industries. Concomitantly, in the following year there was an unprecedented movement of 12 000 people from Quang Ninh to Hong Kong.

Each year, more than 1000 people still drift, of their own accord, to Cam Pha looking for work. Such people, like the government mine recruits before them, are prepared to move hundreds of miles away from the familiar surroundings of their home villages because of the extremely limited prospects of the regions from which they come. Few return and visits are rarely made more than once every one or two years.

The movement of urban workers into Quang Ninh overlies previous population dislocation, following the confrontation with China in 1978–9, when thousands of ethnic Chinese left the Province. In Cam Pha District, approximately 90 per cent of the population left – all Chinese – leaving a residue of lowland Vietnamese (Kinh) people. In place of this population, the government moved in other families from neighbouring districts or provinces. The village of Cam Hai in Cam Pha Town is almost entirely composed of people who were ordered to move from their homes on the Chinese border in 1980. The few scattered Chinese family members who remained usually followed their relatives at later stages: in the late 1980s when it became easier to move across the Chinese border. Unfortunately for them it had then become much more difficult to gain refugee status in Hong Kong.

Altogether approximately 35 000 people, usually the unemployed, resettled on newly-available lands. The process continues. As elsewhere, the aim of provincial authorities has been to distribute resources evenly among the population, and, in particular, incorporate ethnic minorities into mainstream society. In Ba Che district, Tay and Dao communities were moved on to the paddy rice holdings abandoned by ethnic Chinese when they fled across the border. The Dao were swidden cultivators and have experienced some problems in adjustment. Along the coast, fishing families, previously living on their boats, usually on offshore islands, have been resettled in provincial housing and allowed to continue their fishing from the shore. Others, some 600 households, have been moved out to islands, notably the Co To Islands close to Hai Nam, China, and from where several thousand people left for Hong Kong in the 1980s.

Recently, in Cam Pha District several hundred people, relocated there 10 years before, have moved again elsewhere in the same district, because the land originally allocated to them could no longer provide support. Similarly, in the southern coastal district of Yen Hung, shortage of work in the towns has led the District authorities to develop relocation schemes that attempt to match under-developed land to populations that are under-resourced.

Another group in the relocated sector of the population in the Province are the asylum-seeker returnees from Hong Kong who have become a connecting link in channelling international development aid to communities in Quang Ninh. The demographic trends within this Province demonstrate that the returnee experience of seeking opportunities and getting re-established within the community is not an unfamiliar pattern within the population as a whole.

ASYLUM SEEKERS AND RETURNEES

In the following discussion of asylum seekers, those who flee Vietnam on political grounds are not considered; their situation is examined elsewhere.[17] Instead, the focus is on those migrants, at least 60 per cent of the total, who are responding to economic policies and conditions in Vietnam. The province of Quang Ninh is a particularly suitable location in which to examine this focus, because it is an easy exit point for Hong Kong.

The viability of external migration as a workable and successful strategy is self-evident in many northern coastal settlements. For example, in a village close to Hai Phong, there is an extravagantly-decorated newly-built house. It is the home of grandparents, who have always led impoverished lives supporting their family of seven with odd jobs and fishing. Their situation changed when the youngest son, who left for Hong Kong in 1986, was accepted as a refugee and made his home in Canada. Since his arrival there, his remittances have ensured a relatively luxurious lifestyle for his relatives.

In a small community, it is not surprising that, following the economic confusion of restructuration in 1987–8, many families decided that their best chance of safeguarding their living and their children's future was to travel to Hong Kong. One young man was typical of many others when he said that he argued for a long time with his parents to persuade them to let him go and try his luck. Eventually, they agreed because they recognized that with the withdrawal of state subsidies there were no jobs available. The son arrived in Hong Kong after the introduction of screening procedures. He stayed nine months before deciding to come home again, driven back by homesickness, the difficult life in the camps and the hopelessness of being accepted as a refugee.

Another fishing family of nine also returned from Hong Kong. They stated that when they had belonged to the fishing co-operative their living had been reasonably secure, since the state paid for everything; but after the withdrawal of subsidies when they began operating on a private basis, things became much worse. They were labouring people, unused to thinking about how to organize things for themselves, and when they heard about the possibilities of making a living abroad, they had a family discussion and decided to go.

It is common for a family in considering its economically unstable position, particularly as it existed in 1988, to come together to decide what is the best action to take. One family member, perhaps a second or youngest son and his family, will be agreed to be the individual who should try his luck abroad. Other sons or daughters need to remain to look after the parents and their living conditions. The expenses of the trip are raised from the family group, perhaps US$250, much as if it was an investment, since if that individual manages to resettle in a third country he will be able to send back enough money to amply compensate the group. The house, boat and other items, furniture and so forth are 'sold', or given to other family members, being especially careful to keep the house in the hands of a friend or relative so that the state does not exert its right to occupy an empty dwelling.

It should perhaps be emphasized that what has been described is a typical, not a universal case. There are plenty of other examples of boys who asked no one if they could go, some who are escaping criminal prosecution, or of impoverished people who were given a free ride to Hong Kong. Generally, we can say that people who have fairly readily agreed to go home tend to move back into strong family networks and take up an economic situation that approaches at least an average condition in the community.

Nearly 3000 Vietnamese have returned voluntarily to Quang Ninh, and in June 1991 I conducted a small pilot study which included 50 interviews with returnees, scattered in contrasting communities throughout the province. Thirty-six of the interviewees fell within the 18–35 age group, 16 of the sample were women, and all but six belonged to extensive family networks, defined as more than three interrelating nuclear groups. Just over half the sample (28) reported the presence of relatives abroad or in Hong Kong. Eighteen respondents had been relocated in the last 12 years from other regions of the Province, or had parents who had migrated to Quang Ninh looking for work. Half of those of working age before departure had done some skilled work. Although everyone spoke of hard economic times with increasing difficulties in 1988, there was no clear indication of sudden extreme hardship at that time, which might have forced this group of people to leave, but rather a situation of ongoing problems that promised no foreseeable hope of future change. The majority agreed that livelihoods were more stable before the withdrawal of state subsidies, although their real incomes might not have changed very much after the end of subsidization. The security of working for the state with all the benefits of pensions and medical aid disappeared for those who could no longer find employment in the state sector. On return, trade became the main means of earning a living.

Overall, with two exceptions, the returnees interviewed in the urban areas of Quang Ninh were not impoverished; they occupied reasonably-equipped block-built homes, owned either by themselves or their families. Conditions in the rural areas were more restricted, and characterized by occasional employment in agriculture and fishing. Two of the villages visited were impoverished, but returnee families once again reflected an average level of income in the community, not the poorest. The habit of migration was deeply entrenched; in one village on the coast of Quang Ninh more than a third of the community had relatives abroad. Attachment to the village location was thinly developed in all rural areas studied, most families having been resettled in recent years.

An indication of how representative of the province's population, both in attitude and circumstances, returnees tend to be, is demonstrated by the general tolerance and often warmth with which they are taken back into the community. There may be local disagreements about distribution of goods or accommodation, but on the whole everyone understands why returnees went and why they had to come back. Other residents have considered the option themselves and many have relatives who have been successful in settling abroad. Eastern Europe was once a common alternative for some of the people of Quang Ninh, since mining is a high-status occupation, formerly favoured in foreign recruitment programmes. That option is now closed.

CONCLUSION

Migration is a social process that is historically situated in a politico-economic context. The purpose of the analysis has been to explore some aspects of this context, particularly in locating the impetus for spontaneous migration, and pointing out what are at least coincidental linkages between the occurrence of internal and external migration, related to lack of economic development. In Quang Ninh, these connections are direct and almost immediate. Elsewhere, in view of the scale of relocation, such connections remain unproved, but nevertheless it is certain that there has developed a habit of movement not normally associated with the traditional image of villages in Vietnam. A survey recently conducted by the government Department for the Reallocation of Population and Manpower found that once a person had moved the more likely s/he was to move again.

The removal of government restrictions on movement coupled with a reduction in government support has tended to increase the numbers of skilled people moving from unstable situations. Not surprisingly, the more destitute the family the less likely they are now to consider moving, except for short distances when they may return to family in the place of origin. Since 1987, uncertainty about the future is likely to provide a powerful motivation for families able to make a living, but who fear increasing destabilization.

Probably, there exists the feeling that the migration option has the tacit support of the state. Whereas the migration of refugees may be seen as a protest against a government ideology, the movement of economic migrants from Vietnam is in essence implicitly obedient and responsive to state directives, both in their weak and strong forms: the state, either by default, allows people to make their own moves, or else, in its strong form, incites or demands moves for reasons of national interest. It is of course illegal to leave the country without permission and strenuous official efforts are made to prevent such movements. However, after 1987, the doctrine of non-dependency on the state and the failure of the state either to stabilize livelihoods in the New Economic Zones or provide job security in industry, encouraged many would-be migrants, particularly from north Vietnam, to justify to themselves the option of migrating abroad as an acceptable alternative to making a living in Vietnam.

NOTES AND REFERENCES

1. L.A. Wiesner, *Victims and Survivors: Displaced Persons and Other War Victims in Vietnam, 1954–1975*, (New York: Greenwood Press, 1988), p. 345.
2. T. Rambo, 'The Dynamics of Refugee Migration in Vietnam', MA thesis, 1969, University Microfilms Inc., Ann Arbor, Michigan.
3. N. Thrift and D. Forbes, *The Price of War: Urbanisation in Vietnam, 1954–1985* (London: Allen & Unwin, 1986), p. 93.
4. Ministry of Labour, Invalids and Social Affairs, (MLISA), Dept for the Reallocation of Population and Manpower, *Highlights of the Organised Migration Programme in Vietnam* (Hanoi, undated); Communist Party of Vietnam, (CPV), *Draft Strategy of Socio-economic Stabilization and Development of our Country up to the Year 2000* (Hanoi: 1990).
5. W.J. Duiker, *Vietnam since the Fall of Saigon* (Athens, Ohio: 1989, Southeast Asia Series, no. 56A), p. 99.
6. L.M. Hitchcox, 'Vietnamese Asylum Seekers in Hong Kong: An Overview of Fieldwork in the Detention Centres of Hong Kong', unpublished report, 1990.
7. A. Fforde and S. de Vylder, *Vietnam – An Economy in Transition*, Swedish International Development Authority, June 1988.
8. For example, see A. Fforde, 'The Political Economy of Reform in Vietnam: Some Reflections', paper given at Indochina Seminar Series, Harvard Institute for International Development, March 1991.
9. If the government can move from its traditional role of central planning, strengthening its authority over a multi-sectored economy as it does so, there is much that it can achieve, particularly in providing the legislative and

administrative infrastructure for foreign investment and trade. It is, of course, quite possible for a market system to develop within the authority of a one-party state, as several other countries in South-East Asia can bear witness.

10. C.A. Thayer, 'Building Socialism: South Vietnam since the Fall of Saigon', Research Paper no. 20, Griffith University School of Modern Asian Studies, Centre for the Study of Australian Asian Relations, 1982.
11. See MLISA, *Highlights* for further details.
12. Additionally, during the period 1981 to 1990 the government organized the resettlement of 17 048 people from Hanoi, 837 from Haiphong, 26 919 from Danang, 34 938 from Ho Chi Minh, 96 947 from Hue and 34 522 from Nghe Tinh.
13. As a rough comparison in Indonesia it costs approximately $5500 to move a family.
14. As of September 1991, 10 000 dong was equivalent to one US dollar.
15. MLISA, *Highlights.*
16. W.R. Geddes, *Migrants of the Mountains: the Cultural Ecology of the Blue Miao of Thailand*, (Oxford: Clarendon Press, 1976).
17. L.M. Hitchcox, *Vietnamese Refugees in Southeast Asian Camps* (London: Macmillan, 1990).

10 The Movement of Population to the West of China:

Tibet and Qinghai

Graham E. Clarke

INTRODUCTION

This chapter gives a descriptive and historical account of population movements in Tibetan areas in the polity of China, in particular of the present Tibet Autonomous Region (TAR) and Qinghai. It covers related demographic, ethnic, political and economic issues, and in particular concerns the linkage of local communities to wider, dominant, state structures.

The main ethnic group or people within the polity of China are the Han Chinese, who form approximately 92 per cent of the population of some 1200 million;[1] yet in the west of China it is non-Han peoples who, historically, have been dominant both demographically, and to a degree politically, well into the twentieth century. Given the low numbers overall relative to the Han majority, if these non-Han peoples were distributed evenly over the polity they might not be a political force or issue, but ethnic difference is matched by an equally basic geopolitical and geo-economic distinction between the eastern lowlands and western highlands of China. Nationalities other than Han traditionally have occupied nearly half the land area to the west, which is an upland plateau region with a very low population density. Though agriculture is as important in terms of economic value, transhumant pastoralism predominates in terms of area of land given over to that economic form and in this sense is of great general political significance.[2] The Tibet Autonomous Region, with only 4.5 per cent to approximately 17.5 per cent non-Tibetan population is still predominantly pastoral and non-Han.[3]

Culturally exotic people often will be found in the border regions of empires. In modern times borders are no longer the temporary edge of empires expanding against nature, remote from the metropolitan centre, but

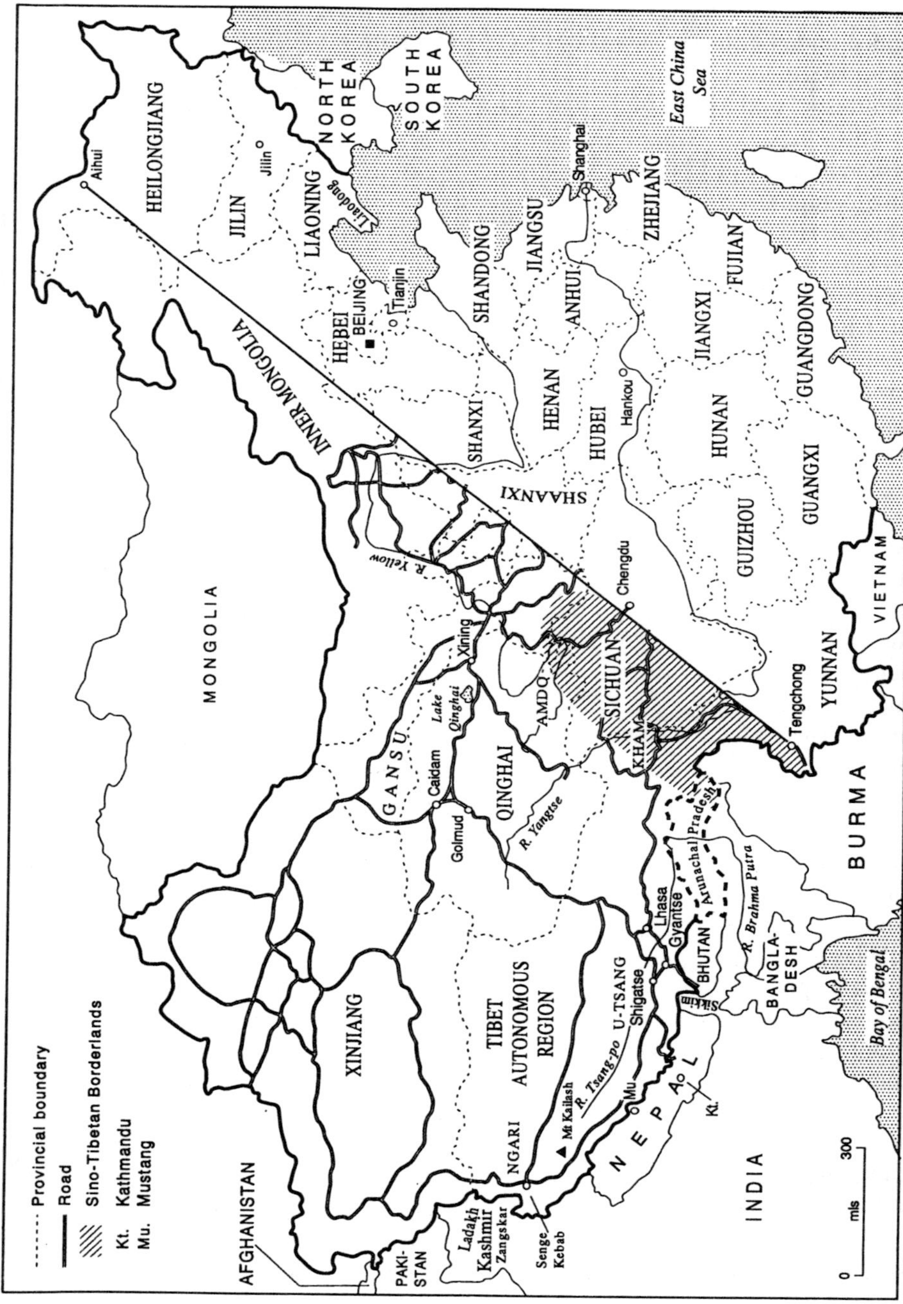

Map 10.1 China

form politically sensitive international boundaries. Hence they have an added importance. The incorporation of the culturally differing peoples of these areas within a bureaucratic state presents political problems, both of control and of the basis of citizenship.

This chapter contains descriptive material from four seasons of field research, 1989 census data and secondary sources on the demographic distribution of ethnic groups, in particular of Tibetans, in western China. This is work in progress, and although the interpretation of the data could be refined and in places possibly corrected, by further field research, it is advanced enough for publication.[4] It may be the first presentation outside of China on the ethnic distribution of peoples based on systematic observation in the region rather than from observations that may be biased towards samples close to roads and urban areas;[5] it also presents census data disaggregated by nationality at the 'county' level for the provincial level units of the Tibet Autonomous Region and Qinghai.[6]

Though reliable in the standard statistical sense of replicable there are problems with the interpretation of such census data.[7] The definition of who is a resident is one such major issue. Chinese population registration data exclude armed forces personnel, travellers who would be passing through in hotels, and some registered as resident elsewhere.[8] In practice, the survey system often is based on a registration by the local administration rather than a survey by a special survey team, and is successively compiled at township, county, prefecture and provincial levels. There are variations in reporting practice from place to place, and the technical inadequacy of such data is especially marked in an area with few administrative resources.

There are other problems in presentation. The names of county level units and peoples vary according to source, and Romanizations of Tibetan and Chinese can follow a number of systems. Some are rough orthographic equivalents to aid approximate pronunciation; others are more precise scholastic transcriptions from either Tibetan or Chinese. Customary English transcriptions of Tibetan names, Romanization of Sinicized equivalents of Tibetan, lack of consistency in official sources, and the culturally normative aspect of naming itself, make the matter even more complex. We have not solved this problem here but have given alternatives in the text where confusion could arise.[9] Well-known English equivalents of names are used on the county Map 10.2, for example, we have Shigatse rather than Xigaze or Rikaxa, and Pa-nam rather than Bainang or Bailang. In the text we follow the practice of this map.[10]

The term Tibet itself is used to refer to areas with different boundaries. The potential for miscommunication is large, as the borders of the topographical area of the Tibetan Plateau, the Tibet of Western travellers up to modern times, the Tibetan Autonomous Region of the People's Republic of China, the areas where Tibetan nationals reside, and the region where Tibetan culture is dominant, are not all the same thing. The term Tibet now may be used by Chinese scholars to refer in a general way to the wider Tibetan cultural area. However, in common practice and administrative Chinese convention the term Tibet implies the Tibet Autonomous Region (here abbreviated to TAR): this refers solely to central, central southern and western parts of this wider cultural Tibet, that is central and central-southern Tibet known together as Ü-Tsang, and the western area of Ngari (Sinicized as A-li)[11]; but it does not refer to the contiguous eastern regions in western Qinghai and Sichuan where half the Tibetan nationals in China live.[12] In many earlier Western sources these areas are referred to as Kham (the far northern areas of Yunnan Province, far western Sichuan Province and a small part of the south of Qinghai) and, to the north of Kham, Amdo (being most of Qinghai Province, part of Gansu Province and a small part of north-western Sichuan Province).[13]

EARLY INCORPORATION AND SETTLEMENT IN THE WEST OF CHINA

There is a turbulent history of military relations between China, the early Uygurs, Turkic speaking peoples and Tibetans attested back to at least the seventh century from Tibetan sources. At that period peoples from the pastoral highlands could dominate the agricultural peoples of the lowlands. The assumption of power of the Yuan Dynasty and the history of other pastoral peoples illustrates this point. For example, when Tibet dominated in the eighth century, edicts enforced the wearing of Tibetan clothes, the celebration of Buddhist festivals, and the cutting of pigtails by the Han Chinese in an area that is a part of modern Gansu.[14]

From the twelfth until the seventeenth century, repeated Mongolian military incursions from the north-east and north-west of China led to settlements near Lake Ko-ko Nor, in the region that subsequently became known as Qinghai.[15] A Qing army also had passed through central Tibet, and reached Lhasa in 1720. Chinese forces subsequently came to central Tibet both to escort their political envoys to Lhasa, and to repel two invasions of Tibet by Nepal; in the first of these in 1792 Chinese forces descended through the Himalaya and approached within half a day's walk of Kathmandu.[16] Yet there was no attempt at settlement or the direct incorporation of central Tibet by the Chinese state at this time.

North-eastern Qinghai was incorporated within China well before central Tibet. After the victory in Xining in 1723, the Qing army began to consolidate its position in the north-eastern lowlands, and there were problems in the control of the wider area in 1822. Qing fought Tibetan forces who had crossed from the south bank of the Yellow River northwards to reach traditional grazing lands. Later, various measures of control were attempted which included restrictions on freedom of movement, trade encouraged through Hui Muslim and Han merchants who entered the area, taxation, and the setting of Mongol and Tibetan against each other.[17]

This did not correspond to administration, but more to a symbolic incorporation and delegation of power. In the territory to the south the Qing government tried to extend their authority, as had the Ming Dynasty in times before, by the various symbolic devices of awarding banners, and by rewarding local leaders who forwarded what could be interpreted as tribute; they also listed, named, and re-classified local clans and counties in their records. Such symbolic incorporation was reinforced by occasional military forays which had mixed results. There are clear records of Han peoples in north-eastern Qinghai in the late nineteenth century.[18] There is little to suggest that Qing forces in practice controlled much land beyond a corridor north of the Yellow River and eastward from Xining to Gansu; nor is there much evidence of Han settlement in Qinghai to the south of the northern bend of the Yellow River near Xining. Yet this minimal control marks the beginning of large-scale population movements by Han peoples from lowland China, people who carried out agriculture and who began to move upwards to settle in the lower north-east of Qinghai.[19]

POPULATION PRESSURE AND RESETTLEMENT IN THE QING PERIOD

After some fluctuations in the seventeenth century, in the first quarter of the eighteenth the population of China probably rose from 100 to 150 million. By 1795 the Chinese population was probably over 300 million, and by 1850 some 430 million; growth between 1749 and 1851 averaged 0.9 per cent per annum.[20] A growing population meant increased pressure on agricultural resources, which in turn led to population movements.

Despite Imperial edicts against leaving the territory, there was a large-scale movement to South-East Asia under a pull from Dutch and British colonial expansion. There was also large-scale migration to the eastern seaboard of the Americas, which included 100 000 migrants to Peru alone in the nineteenth century; in the USA there also was a pull both from the

California Gold Rush of 1848–9, and the railway construction industry. These Han overseas migrants tended to work as farmers, labourers, or local merchants. Many at first were single men, abroad to make their fortune with the intention of returning with wealth to their homeland and their families; they tended to work and stay together according to local regional groups and dialects. For indentured labourers, there was more than an element of compulsion to staying abroad, and return was not always easy. Many, later, became permanent settlers, and were joined by women from inland China. There are possible parallels here with twentieth century movements into western China.

Overall, old lands were resettled and new lands were opened up, and the period is marked by a rapidly expanding population and movements across the whole countryside.[21] Within China itself there were coastal and internal population movements north-east to Liaodong, Jilin and Heilongjiang, and to expanding cities such as Hankou, Shanghai and Tianjin. Even during the seventeenth century, pressure for land had already moved Han settlers south-westwards, and settlement in Yunnan in the south-west continued the trend established under the Ming Dynasty. State policy was to establish military garrisons in remote areas. Ideally, these would be self-supporting, which also has clear continuities with current practice.

The state promoted agriculture, to which end civilians were encouraged to establish colonies in remote garrisoned areas, as well as to resettle internally in other areas depopulated by civil war. The scale of movements and depredations, Han against Han as well as against non-Han nationalities, was massive, and resulted in the depopulation of whole areas of agricultural land which then had to be resettled. For example, civilian deaths in China from 1851 to 1864 during the Taiping rebellion have been assessed at up to 47 million adults.[22] Han farmers both suffered these depredations, and carried out much of this resettlement and new cultivation; if an indigenous people in an area dissented, it was put down severely. The Miao population of highland Hunan was displaced at this time, and deforestation, terracing and cultivation were the rule in these areas.

Besides the early Qing Imperial successes and the increase in population, another factor led to a change in the eighteenth century, the availability of New World crops, which provided the means to resettle and colonize hill areas. Besides high-yielding rice on the irrigated plains there now was also maize, sweet potato, peanuts and the potato, all of which could be grown on the largely rain-fed, and in places terraced, hillsides. These crops spread in the early nineteenth century in Sichuan and Yunnan,[23] as well as on hill-land in the south-east of China in Jiangxi, Fujian, Hunan, Zhejiang and southern Anhui provinces.

Initially, new cultivation by slash and burn and terracing increased yields; but there were major problems from soil exhaustion and erosion, with subsequent siltation and flooding. This was especially the case with maize and tea cultivation.[24] As hillsides were progressively exploited a continued influx of population resulted in declining absolute as well as *per capita* yields. In the lower Yangtse, there were even laws passed forbidding the renewal of leases to tenant farmers so as to maintain production levels. Hence there was a continued pressure to move outwards and upwards, from the agricultural basins and valley-bottoms of lowland China to other regions further upstream towards the highlands, the west of China, where the hills formerly were a barrier.

In the north-east of Qinghai this new agriculture was capable of producing a surplus, which could be used to finance the military garrison in Xining on whom Han settlers relied for protection from both Mongol and Tibetan horseman. This surplus permitted an efficient specialization between army and farmer which helped to maintain and expand the area under state control in a way formerly not possible. Thus, the balance of peoples on the hill-land just off the eastern Tibet plateau began to be reformed in the late eighteenth century, as Han peoples moved up the valley to hill tribal areas with New World crops, but below the Tibetans who were higher still on the plateau.

There was no settlement of Han peoples on the Tibet plateau itself where pastoralism remained the dominant economic form. Hill-land between the edge of the plains and the plateau could be settled; but in going westwards or north-west, farmers still had to stop at the high Tibetan plateau (or at the deserts of Gansu). In many areas the altitude was above the limit for the production even of New World crops; moreover, the establishment of successful military garrisons was more difficult in a high mountain and nomadic area than in an agricultural and sedentary area. There was a limited political presence further west in central Tibet and elsewhere on the highland plateau, more symbolic than directly administrative, and maintained only by occasional military expeditions.

In Qinghai, by 1860, an approximate boundary was generally recognized between land administered by the Qing state and that controlled by nomads. The 'sun and moon' ridge, together with the trade market at Hongyuan, west of Xining and east of Lake Qinghai, was the line between highland pastoral and agricultural areas until 1949. Tibetan peoples were to the south and west; some Mongolian peoples to the far west and north; Han and other peoples to the north-east. Even in the first half of the twentieth century under Nationalist China there was little further administrative control. Until the advent of modern technology, and the establishment of the People's Republic of China (PRC) in 1949, administrative control and population movements westwards were halted by the Tibetan plateau itself. Away from the north-eastern area of Qinghai, borders were, according to Rock who travelled in eastern Tibet at that time, 'ill-defined and more or less arbitrary'; the state was not even able to send postal couriers through the area of southern Qinghai, that is Golog and Yushu.[25]

POLITICAL ORDER IN THE BORDERLANDS

Both this area of Qinghai and west Sichuan (Amdo and Kham) were of Tibetan culture, with institutional links to Lhasa through the Buddhist religious hierarchy. Their internal political order centred on changing local territorial units linked by kinship,[26] and even after a nominal political unification in the nineteenth century one such dominion in Kham was still known as the '80 000 families' of Nyarong.[27] Little secular political authority linked this traditional borderland to the outside, and Lhasa was played off against Beijing to help maintain local autonomy.

This changed in the nineteenth century. Under the prompting of the Qing government Lhasa formally took over this eastern area of Tibet, and in 1865 began to receive tribute in the traditional manner. In this period the civil authority of Lhasa also extended to much of Golog and Yushu, regions subsequently transferred to Qinghai, and Beijing recognized the local authority of Tibet over a wide area of Kham that centred on Nyarong in present-day western Sichuan.[28] However, in 1896, the Qing government sent their own army to this frontier of Sichuan and began to incorporate it directly into their own western administration.

The cruelty of military rule against civilians there was no different from that in lowland China, or from that of earlier periods of Chinese history. Yet in practice, warlords were not as effective against armed tribesman on horseback with movable livestock in these western highlands, as they continued to be against settled agriculturalists reliant on fixed, seasonal, cultivation in the lowlands. By the beginning of the twentieth century the Lhasa government supported the local powers in Kham against these Han troops, and by 1912 the Qing empire had collapsed. Subsequently, the early twentieth-century history of conflict and of Tibetan opposition to Nationalist Chinese forces is an important feature in the formation of the modern nation. A common Tibetan sentiment and identity developed in reaction to this external force, out of the more traditional and fluid local political and religious identities in a common cultural milieux.

The Nationalist forces disintegrated in Kham, and were forced out of the north by Tibetan nomads of Golog.[29] The PLA entered Tibetan areas of Gansu and then Qinghai in the summer of 1949; they entered Kham in 1950 in the aftermath of the disintegration of the Nationalist forces, and in contrast to what passed before them they were at that time a well-behaved, civilized and orderly force. They proceeded westwards to Lhasa, where they arrived in 1951. Telegraphic communication was a key factor in the establishing of control in the area, as it had been for the British forces which had entered Lhasa earlier in 1904. Yet the establishment of control still was not straightforward on this mountain terrain: there was armed resistance to the Chinese military in 1952–3, and subsequently. Though day-to-day control of main routes in the south of Qinghai largely followed the construction of roads starting in the 1950s, sporadic armed resistance still continued.

The equilibrium between the pastoral and new agricultural areas that had been established during the Qing Dynasty began to alter again, as the modern industrial state extended westward. It did so through the construction of airfields and roads, bureaucratic registration, and the incorporation of communities through civil political campaigns. These campaigns were carried out with a vision of material modernity, progress, and universal socialism according to the model developed earlier in lowland China. Though the military presence remains the final authority of the state, the period to date is one of social and political incorporation of the plateau in which material technology and economic forces play a major role in initiating change, supplanting the largely ceremonial and military interventions of earlier times. Here the construction of roads was important not just for the original purpose of military control, but subsequently, post-1978, in promoting market-led economic changes which have a wide impact on the plateau area. In turn these changes have helped to define the vector of such population movements as have taken place, which have tended to follow the roads and the towns.[30] These population movements have led to a further blurring of the former regional and vertical ethnic separation in the lower, eastern, areas of the Sino-Tibetan borderlands, where there were further population movements of Han farming peoples from the late 1950s and early 1960s.

THE POPULATION DISTRIBUTION OF MODERN QINGHAI

In the north-east of Qinghai where currently some two-thirds of the province's population lives, the land gradually slopes down from the highland plateau towards lowland China.[31] This area is predominantly agricultural whereas the central, western and southern areas of Qinghai are still mainly pastoral, with transhumant grazing over extensive grasslands on increasingly sparse pasture that to the west and south ends in desert. Approximately some 80 000 households, that is 900 000 of the total population of around four and a quarter million, still live by raising livestock.

Absolute population totals for nationality by prefecture are presented for Qinghai in Table 10.1; this population includes Tibetan, Mongolian and other nationalities besides the majority Han population. Table 10.2 gives percentages of the populations of each prefecture by nationality. The figures indicate that only some 20 per cent of the population of Qinghai is Tibetan; but read in conjunction with Map 10.2 we can see that they have an approximately even density of population over all the inhabited areas including north-eastern Qinghai.[32] In more detail from north to south, Haidong Prefecture, in the far north-east, is under 10 per cent Tibetan; Haibei Prefecture is under 20 per cent Tibetan; Huangnan and Hainan Prefectures are over 50 per cent Tibetan; and Golog and Yushu are around 90 per cent Tibetan: that is, the same percentage as in the adjacent rural areas of the Tibet Autonomous Region. Overall, the percentage Tibetans form of the population increases to the south-east, to the prefecture of Golog bordering on Gansu and Sichuan, and to the south-west, to the prefecture of Yushu that borders on the Tibet Autonomous Region.

Some 60 per cent of the population of Qinghai is Han, and they are concentrated in the lower north-east where some 84 per cent of settled Han live. Subsequently there have been internal resettlements of Han peoples further west in Qinghai, in particular to Haixi. In addition, some 14 per cent of the provincial population are Hui Muslim, who are concentrated largely in the north and east; Mongolian, and the Tu and Salar minority nationalities, form less than four per cent of the total population.[33] The Tu and Salar are highly localized, each in two separate counties of the north-east.[34]

At this lower county level of disaggregation, the Tibetan population, with some exceptions in the far north-west, still has a relatively even absolute distribution: all rural counties have a significant Tibetan population, nearly three-quarters of them have more than 10 000 Tibetans, and the smallest counties have the lowest absolute numbers of Tibetans.[35] Overall, the smaller counties in the north-east tend to be lower in altitude and are those where the Han population is concentrated; some contain urban areas. Han are also found further west in the agricultural and urban counties of Qinghai.

A relatively uniform low density distribution is to be expected, *a priori*, from a pastoral population dependent economically on transhumance over open grassland. The historical pattern suggested is of the superimposition

Table 10.1 Qinghai Province: Absolute totals for nationality by prefecture

	Han	*Tibetan*	*Hui*	*Mongol*	*Other non-Han*	*Non-Han sub-total*	*Total people*
Municipal areas	770 857	25 769	165 056	4 315	36 745	231 885	1 002 742
Haidong Prefecture	1 211 261	154 494	313 733	6 442	160 081	634 750	1 846 011
Haibei Autonomous Prefecture	120 438	45 513	62 488	10 343	7 149	125 493	245 931
Huangnan Autonomous Prefecture	15 896	107 997	13 404	21 409	6 588	149 398	165 294
Hainan Autonomous Prefecture	146 777	176 622	20 447	2 169	3 689	202 927	349 704
Golog Autonomous Prefecture	11 277	97 449	702	69	456	98 676	109 953
Yushu Autonomous Prefecture	8 109	210 316	309	37	209	210 871	218 980
Haixi Autonomous Prefecture	218 773	26 050	8 822	19 498	4 132	58 502	277 275
Total	2 503 388	844 210	584 961	64 282	219 049	1 712 502	4 215 890

Source: Based on author's research and compilations of the Qinghai Government, the China Statistics Publishing House, Tianjin.

Table 10.2 Qinghai Province: Nationality as a percentage of population of prefecture

	Han	*Tibetan*	*Hui*	*Mongol*	*Other non-Han*	*Non-Han sub-total*
Municipal Areas	76.87	2.57	16.46	0.43	3.66	23.13
Haidong Prefecture	65.62	8.37	17.00	0.35	8.67	34.38
Haibei Autonomous Prefecture	48.97	18.51	25.41	4.21	2.91	51.03
Huangnan Autonomous Prefecture	9.62	65.34	8.11	12.95	3.99	90.38
Hainan Autonomous Prefecture	41.97	50.51	5.85	0.62	1.05	58.03
Golog Autonomous Prefecture	10.26	88.63	0.64	0.06	0.41	89.74
Yushu Autonomous Prefecture	3.70	96.04	0.14	0.02	0.10	96.30
Haixi Autonomous Prefecture	78.90	9.40	3.18	7.03	1.49	21.10
Province	59.38	20.02	13.88	1.52	5.20	40.62

of later higher-density settlements of agricultural and urban Han peoples in specific locations, in particular in the lower areas. This demographic picture accords broadly with the historical evidence presented above, and suggests the early, widespread, presence of culturally Tibetan pastoral peoples, together with Mongolian and related non-Buddhist peoples in the entire region, including the north-east near the Yellow River.

POPULATION MOVEMENTS IN QINGHAI AND THE PEOPLE'S REPUBLIC OF CHINA

Military control brought the possibility of a further expansion of agriculture upwards and westwards into these regions of the north-east previously under the control of nomadic peoples. Along with the agriculture, there could be a further settlement in associated administrative and service industries. Further population movement onto the higher parts of the plateau itself, in the south-west, remained limited, because of the extreme topography. Army settlement helped to maintain control of these lines of transport and communication, and such army farms can be seen alongside main highways in strategic areas. These military settlements support themselves economically, much as did the Qing forces before them.[36]

The indigenous Tibetan population now grow grain for market sale on their own land, which in a *de facto* sense can also now be rented out. To the immediate east of the 'sun and moon ridge' there are now Han farmers, tilling this land that previously was pasture. A shift in emphasis in the economic base has occurred, from pastoral subsistence and restricted local trade, to a more mixed agricultural production with market trade.

Agriculture began to move beyond the 'sun and moon' ridge into the pastoral area under collectivization post-1950, and the area under agriculture has increased with the economic liberalization post-1978. Between 1982 and 1989 the area under grain in two rural townships in this area has doubled, with a corresponding 10 per cent drop in numbers of sheep.[37] Here, as elsewhere in western China, reforms in the context of the existing transport network allowed economic growth.[38]

Overall in Qinghai some 467 000 hectares of new agricultural land was registered between 1958 and 1989. This total is small compared with the 40 million hectares of grassland in Qinghai as a whole, used mainly by Tibetan and Mongol pastoralists; but it includes prime irrigated land. Valley land has been reclaimed and settled in the Xining area, and also in new agricultural regions in the basin south of Caidam, south of the road westwards through Mangya town to Xinjiang, north of the city of Golmud in Haixi prefecture (see Map 10.2).[39] Roads and transport, generated by the need for military security, have allowed population movements into the area and increased agricultural production. In turn, increased food production can now sustain an urban population on a more permanent basis than earlier military incursions.

The absolute population figures for Haixi prefecture by county and nationality are given in Table 10.3, and allow a breakdown for Golmud municipality. Golmud was a new town founded and built up in west Qinghai in the mid-1960s, and is the western railhead from Xining and lowland China from where goods have to be transshipped onwards by road to Lhasa and central Tibet. The Mongolians and some of the Hui Muslim too were long-established there, though in all likelihood there has been a recent influx of Hui Muslim along with Han peoples in service trades. Settlement at Golmud started in 1966, and the population was already some 57 202 in 1982. After that time population appears to have grown at only 1.5 per cent per annum up to its 1989 figure of 64 179, indicating a more recent brake to urban migration. As indicated in Table 10.4, in 1989 the Tibetan population approached 4 per cent, the Mongolian was under 3 per cent, and the Han population was around 90 per cent. Clearly, Golmud is a new-town, populated primarily by Han migrants, that has grown up in a sparsely populated remote pastoral area in the west of Qinghai largely prior to 1982.

Overall, Haixi Prefecture is 79 per cent Han, who are found in large numbers in all but Tianjun County. Only some nine per cent of the population of the prefecture is Tibetan and they are registered mainly in counties between Caidam and Lake Qinghai, in particular in Tianjun and Dulan counties, where they form 80 per cent and 18 per cent of the

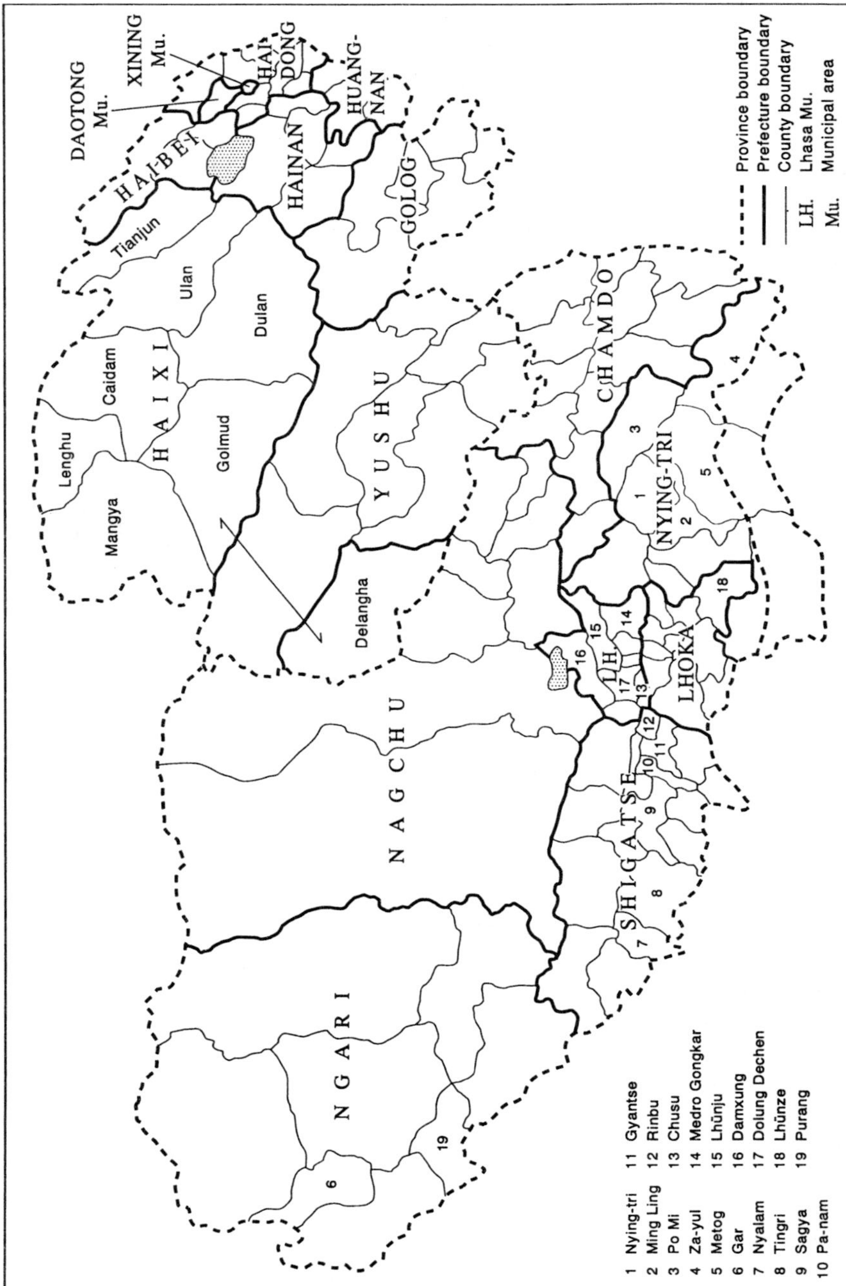

DAOTONG Mu.
XINING Mu.
HAI DONG
HUANG-NAN
HAI-BEI
HAINAN
GOLOG
Tianjun
Ulan
Dulan
HAIXI
Caidam
Lenghu
Golmud
Mangya
YUSHU
Delangha
CHAMDO
NYING-TRI
LH.
LHOKA
NAGCHU
SHIGATSE
NGARI
Province boundary
Prefecture boundary
County boundary
LH. Lhasa Mu.
Mu. Municipal area
1 Nying-tri
2 Ming Ling
3 Po Mi
4 Za-yul
5 Metog
6 Gar
7 Nyalam
8 Tingri
9 Sagya
10 Pa-nam
11 Gyantse
12 Rinbu
13 Chusu
14 Medro Gongkar
15 Lhünju
16 Damxung
17 Dolung Dechen
18 Lhünze
19 Purang

Table 10.3 Qinghai Province, Haixi Prefecture: Ethnic groups by county (1989)

	Han	*Tibetan*	*Hui*	*Mongol*	*Salar*	*Tu*	*Other*	*Total (1989)*	*Total (1982)*
Haixi Autonomous Prefecture									
Golmud Municipality	57 834	2 498	1 545	1 745	54	134	369	64 179	57 202
Ulan	22 264	985	1 881	5 044	589	627	166	31 556	76 083
Delangha	37 166	1 045	1 016	5 597	41	388	167	45 420	
Dulan	32 564	9 105	2 707	6 617	254	241	201	51 689	60 985
Tianjun	2 712	12 052	206	15	6	134	7	15 132	13 926
Caidam	19 924	147	690	101	21	73	220	21 176	21 793
Lenghu (not available)									17 156
Mangya	46 309	218	777	379	34	155	251	48 123	22 501
Total	218 773	26 050	8 822	19 498	999	1 752	1 381	277 275	269 646

Table 10.4 Qinghai Province, Haixi Prefecture: Percentage populations of Prefecture

	Han	*Tibetan*	*Hui*	*Mongol*	*Salar*	*Tu*	*Other*
Haixi Autonomous Prefecture							
Golmud Municipality	90.11	3.89	2.41	2.72	0.08	0.21	0.57
Ulan	70.55	3.12	5.96	15.98	1.87	1.99	0.53
Delangha	81.83	2.30	2.24	12.32	0.09	0.85	0.37
Dulan	63.00	17.61	5.24	12.80	0.49	0.47	0.39
Tianjun	17.92	79.65	1.36	0.10	0.04	0.89	0.05
Caidam	94.09	0.69	3.26	0.48	0.10	0.34	1.04
Lenghu (not available)							
Mangya	96.23	0.45	1.61	0.79	0.07	0.32	0.52
Prefecture	78.90	9.40	3.18	7.03	0.36	0.63	0.50

population respectively.[40] Mongolians are uniformly distributed in Haixi, except in the far west where few are registered. The far west, Mangya Area, along the southern road to Xinjiang, is 96 per cent Han. Clearly, there is an immigrant Han population now resident largely to the west of long-standing Tibetan and Mongolian populations, bordering on Xinjiang.

Large-scale resettlement of Han peoples into Qinghai for the period between 1954 and the mid-1960s is well recognized. Many were peasants who came to settle on newly-irrigated lands in part from the push of agricultural pressure in lowland China, and include migrants from Shaanxi, Shandong and Henan provinces. Others who came less voluntarily included exiled cadres, relocated as a result of the anti-rightist movements of 1957/8, and those sentenced to serve in 'reform through labour' and prison camps in Qinghai and who subsequently were resettled there. It was at this time that 200 000 people settled at the capital, Xining, in eastern Qinghai: that is, around 20 per cent of the current total of the city.[41]

As well as population movements into the province, there have also been increased temporary seasonal movements of Han and non-Han labour within it. In particular, there have been seasonal influxes of Han men from the east of Qinghai to the highland Tibetan areas, looking to supplement their agricultural income. With the introduction of economic reform policies, such movements have been accentuated. For example, in 1989 there was the attraction of gold prospecting in the west of Qinghai, as there had been previously in the 1930s.[42] Another contemporary example comes from the period between 1983 and 1985; namely the increased numbers who came west seasonally to Lake Qinghai for winter fishing, to the point of over-exploitation.[43] The earlier period of decollectivization had allowed direct control over such movements, which under market reform disappeared.

The increase in grain production of the early 1950s in Qinghai was not sufficient to meet the 1989 needs of the population. The province has an annual grain deficit of some 25 000 to 30 000 tons; grain purchase, distribution, and subsidized sales are the largest item of provincial expenditure, in 1988 taking around a quarter of the annual budget of 500 million yuan. Provincial government policy in 1989 was not just to increase economic output in general, but grain production in particular, so as to reduce imports and the need for subsidies for distribution. From 1982 onwards there have been internal population movements from eastern to western Qinghai, from Hainan and Haidong prefectures to new agricultural land in Haixi. Today, there are also resettlement schemes for people from eastern Qinghai on newly-irrigated land in central-eastern Qinghai. In Hainan, above the gorge of the Yellow River, this land is for forestry and grazing, as well as for grain production.[44]

Han and other peoples have not moved to the southern highlands of Qinghai, where grain production is technically difficult. Given recent intra-provincial movement elsewhere in Qinghai, this suggests that in 1989 there was no policy to promote general population movement into Qinghai, or to settle Han peoples in Tibetan regions *per se*. At the same time, economic pressures from the poorest and most crowded region of Qinghai, the north-east, and the readier take-up of grain production by Han peoples over Tibetan, combines with the current priority for economic growth in agricultural output. This tends to push Han farmers and traders westwards.

POPULATION DISTRIBUTION IN THE TIBETAN AUTONOMOUS REGION

The factors behind settlement by Han peoples in the Tibet Autonomous Region (TAR) are similar to those in Qinghai; particular differences are geographic and historical, as less of the TAR is at a low altitude off the plateau, and there was no earlier period of attempt at direct rule or settlement as in Qinghai.

Absolute figures on population by prefecture are presented in Table 10.5, and percentages in Table 10.6. Current official figures state that Tibetan residents form just over 95 per cent of the overall population of the Tibet

Table 10.5 Tibet Autonomous Region: Absolute totals for nationality by prefecture

	Han	*Tibetan*	*Hui*	*Mongol*	*Other non-Han*	*Non-Han sub-total*	*Total people*
Lhasa Municipality	44 939	327 833	2 476	60	660	331 029	375 968
Chamdo Prefecture	7 019	491 608	126	2	1 407	493 143	500 162
Lhoka Prefecture	5 725	274 226	69	4	783	275 082	280 807
Shigatse Prefecture	4 920	542 209	152	19	1 868	544 248	549 168
Nagchu Prefecture	2 961	290 777	58	9	37	290 881	293 842
Ngari Prefecture	1 435	60 179	10	1	14	60 204	61 639
Nying-tri Prefecture	14 218	109 514	96	11	10 585	120 206	134 424
Total	81 217	2 096 346	2 987	106	15 354	2 114 793	2 196 010

Table 10.6 Tibet Autonomous Region: Nationality as a percentage of population of prefecture

	Han	*Tibetan*	*Hui*	*Mongol*	*Other non-Han*	*Non-Han sub-total*
Lhasa Municipality	11.95	87.20	0.66	0.02	0.18	88.05
Chamdo Prefecture	1.40	98.29	0.03	0.00	0.28	98.60
Lhoka Prefecture	2.04	97.66	0.02	0.00	0.28	97.96
Shigatse Prefecture	0.90	98.73	0.03	0.00	0.34	99.10
Nagchu Prefecture	1.01	98.96	0.02	0.00	0.01	98.99
Ngari Prefecture	2.33	97.63	0.02	0.00	0.02	97.67
Nying-tri Prefecture	10.58	81.47	0.07	0.01	7.87	89.42
TAR	3.70	95.46	0.14	0.00	0.70	96.30

Autonomous Region, and Han peoples only some 3.7 per cent. Though Han and Hui Muslim people form a low overall percentage of the population, the distribution of Han is far from uniform, and is accentuated by varying numbers of temporary travellers and armed forces personnel: Han residents centre on urban areas, townships along major roads, and the lower south-eastern region to the north of India. They have a high significance to the modernizing market economy, in trade and service industries.

Overall, there is a greater percentage of Han per prefecture to the east rather than the west of the Tibet Autonomous Region, which correlates with proximity to lowland China and agriculture. Nying-tri (also commonly spelled Linji) prefecture, a region usually known in Tibetan as Kong-po, is the most significant of these eastern rural areas in terms of the absolute numbers of Han people.[45] Absolute figures on population by prefecture are presented in Table 10.7, and percentages in Table 10.8. The census figures for residence still state that some 81.5 per cent of the registered population of this prefecture is Tibetan, that Han form some 10.6 per cent of the population, and that the balance is constituted by indigenous peoples linguistically related to Tibetans (7.9 per cent).

The majority of the Han peoples are located in the central county of Nying-tri itself, where 28.6 per cent of the registered population are Han, and in two counties in the southern area close to the road to Sichuan, namely Ming Ling and Po Mi – where 10.5 per cent and 9.9 per cent of the registered population respectively are Han.[46] Nying-tri county has early wool-mills, and Ming Ling county has lumber-mills, which may well date

Table 10.7 Tibet Autonomous Region, Nying-tri Prefecture: Absolute totals

	Han	*Tibetan*	*Hui*	*Mongol*	*Other non-Han*	*Non-Han sub-total*	*All people*	*All people (1982)*
Nying-tri	8 634	20 678	74	7	764	21 523	30 157	29 292
Ming Ling	1 574	12 507	4		930	13 441	15 015	13 142
Nang-Dzong	244	11 837	2		14	11 853	12 097	10 920
Kong-po Gyamda	522	20 298	3	2	16	20 319	20 841	18 762
Po Mi	2 460	22 273	13	1	51	22 338	24 798	24 910
Za-yul	705	20 486		1	1 610	22 097	22 802	21 682
Metog	79	1 435			7 200	8 635	8 714	7 642
Total	14 218	109 514	96	11	10 585	120 206	134 424	126 350

Table 10.8 Tibet Autonomous Region, Nying-tri Prefecture: Percentages of prefecture population

	Han	*Tibetan*	*Hui*	*Mongol*	*Other non-Han*	*Non-Han sub-total*
Nying-tri	28.6	68.6	0.2		2.5	71.3
Ming Ling	10.5	83.3			6.2	89.5
Nang-Dzong	2.0	97.9			0.1	98.0
Kong-po Gyamda	2.5	97.4			0.1	97.5
Po Mi	9.9	89.8	0.1		0.2	90.1
Za-yul	3.1	89.8			7.1	96.0
Metog	0.9	16.5			82.6	99.1
Prefecture	10.58	81.47	0.07	0.01	7.87	89.42

from early attempts at industrialization in the 1960s.[47] There are also early reports of settled farming communities of Han in this area of the south-east of Tibet, that is near the roads north of the Indian border; as communes they appear to date from before the formal setting up of the Tibet Autonomous Region in 1965, or slightly earlier at the time of the border war with India in 1962.[48] Hence the majority of the rural Han population probably settled in the 1960s; other Han in the central area have probably come as a result of economic growth in the main town, following both state-owned industry and the more recent market reform.

There is another large non-Tibetan population located mainly in the south of this prefecture, up to and over Za-yul and Metog counties.[49] On Chinese maps Metog in particular descends south through the Himalaya from the northern McMahon line recognized by India, to the old 'outer' line at the foothills of Assam recognized by China, this being the disputed border with Arunachal Pradesh in India.[50] The military conflict between these two powers here in 1962[51] has subsequently become a stand-off.[52] Some 82.6 per cent of the minority peoples of Metog are *other* than Tibetan; Za-yul has a 7.1 per cent minority non-Tibetan population.[53] Chinese statistical records, which here are likely to be in part estimates, refer to these minority populations as Mönba or Löba. These two groups are also resident in the main county area of Nying-tri, presumably in the urban area. They are indigenous, organized by kinship and local village association, speak Tibeto-Burman languages, and are similar to others in adjacent areas of Arunachal Pradesh in north-east India.[54] Their inclusion here as 'non-Tibetans' indicates the Chinese claims to sovereignty over territory south of the Tibet plateau with indigenous peoples as minorities, not a movement of Chinese or any other peoples northwards into Tibet.[55]

The far-western prefecture of Ngari (A-li) is essentially similar to the east in terms of settlement pattern, though there is a lower overall population density and there are few indigenous tribal peoples. Senge Kebab is a new town that, like Golmud, was built up after 1956 in the far-west, in Gar County.[56] It is close both to the border with India in the west, and to the border north to the province of Xinjiang where there is an Uygur population near this area.[57] The figures indicate that Han peoples form 12 per cent of the registered residents of Gar county, and that Tibetans are 87.5 per cent of the county and 97.6 per cent of the prefecture.[58] Informal estimates place the total number of Han peoples at around 4000–5000, that is around 30–50 per cent of the urban population; there are also said to be Uygur present.

Unlike Haixi in western Qinghai, there are few reports of Han rural farming settlements. Hans concentrate in urban areas and close to roads, and informal reports again suggest that these numbers have increased since economic reform in the early 1980s, with more Han and Hui Muslim from adjacent lowland regions of China such as Sichuan. Again, though small in absolute numbers, they have a high profile in the modern service-oriented and private economy, as shopkeepers, building-workers, restaurant owners, and in other industries. The overall pattern, like that of Qinghai, essentially is of a new town founded in the 1960s close to a strategic border, populated by a large percentage of Han, and of Han settlements latterly linked to a market economy in service industries close to roads.

A third illustration of modern settlement pattern in the TAR is given by the population distribution in the central southern prefecture of Tibet, Shigatse, close to the border with Nepal. The total registered population in the prefecture in 1989 is 542 209, of which 98.7 per cent is given as Tibetan; some 96.1 per cent of the 77 397 population of Shigatse, Tibet's second city, is recorded as Tibetan and 2946 are registered as Han; in addition, some 3200 people were registered on one day in the summer of 1992 as resident overnight while visiting the prefecture: informal estimates of numbers of Han people present, which may or may not include armed forces personnel, are higher.[59]

The PLA passed through this area as they did through most rural areas, and as did Qing and Mongolian armies more than once before them. There are military outposts by roads, such as at Tingri and at Shigatse itself. However, though Shigatse town has grown and in 1991/2 Gyantse (the third traditional city of central Tibet) was growing, there are no large *new* towns in this area. Though Pa-nam and Gyantse counties are the main agricultural areas of Tibet, and there is irrigated agriculture in the south close to the Himalaya and the Nepalese border, there is little or no evidence of Han agricultural settlements. In Pa-nam county, which straddles the main highway half-an-hour's drive out from Shigatse towards Gyantse, some 60 Han were recorded out of a population of 37 753. One group there ran a restaurant and, as most Han in the county, they were kin to the technicians, doctors, teachers and accountants at the county headquarters. In Pa-nam county in 1992 there were no Han farmers, and even the sight of Han personnel away from the county headquarters or main road was quite unusual. This is a confirmation of the experience in most other rural areas in central Tibet, where Han or any other agricultural settlers appear not to be found away from the peri-urban areas.[60]

As noted above, there is an earlier history of warfare between Nepal and Tibet in this central southern region. Until 1971, Indian Army troops controlled northern region checkposts in Nepal, and until even more recently Tibetans used northern Nepal as a base for guerrilla attacks into Tibet. In 1974, under Chinese pressure, the Nepalese removed the last of the Tibetans from the border areas of Mustang and Manang.[61] Nepal has received aid from China since 1956, does not question China's sovereignty in Tibet, and since 1960 has had a border treaty with China.[62] It now represents little perceived threat to China, which suggests that the settlements further east and west in Tibet may have been founded as strategic border strongholds in relation to India; that is, military garrison towns originally placed for control at the extreme corners of the Empire.

Between 1951 and 1959, the PLA's use of force to quell disturbances probably led to large reductions in the Tibetan population.[63] The figures presented vary from the thousands to the millions, and it is difficult to be accurate within even an order of magnitude on an estimate of the number of deaths in Tibet at this time.[64] The Cultural Revolution which began in 1966, shortly after the establishment of the Tibet Autonomous Region, was not a period of major migration into central Tibet; rather it was one of general chaos in the face of food shortages, and the maintenance of order in some regions by core-state structures of control such as the army.[65] Deaths and malnutrition also resulted from food shortages during the period between 1966 and 1978.[66]

There are parallels with the suffering in other areas within China for this period and earlier during the Great Leap Forward. The hardship and suffering of the Tibetan people was recognized after the visit of Hu Yaobang and Wang Li in 1980 which, in a directive from the Central Committee of the Chinese Communist Party in Beijing that even used the term 'Han chauvinism' (*Da Hanzuzhuyi*), resulted in administrative reforms for the TAR and other Tibetan areas of western China.[67]

REGULATIONS, AND ECONOMIC INCENTIVES ON SETTLEMENT

The economic and administrative reforms in Tibet and minority areas of Qinghai and western Sichuan that resulted from the Central Committee directive included increases in the local power of provinces, prefectures and counties classified as autonomous, and the provision for their administrative heads to be of the local nationality. At the same time the regional government has to administer the economy under national-level plans and follow national-level directives.

The initial direct effect of the administrative reforms was to reduce the number of Han.[68] However, new economic freedoms were to increase population mobility, in two ways: first, people acquired a right to move around and trade privately; second, access to the extensive road network originally built for security purposes promoted mobility. Combined with the policy of increasing trade and marketing outside of the state sector, these factors resulted in rapid economic growth, and led to movements of peoples in Sichuan, Qinghai and the Tibetan Autonomous Region. This was in part a reversion to the traditional economy, which also emphasized long-distance trade across ecological zones, but the use of large numbers of private trucks, moving in a large arc north from Chengdu and Gansu, west through Qinghai and down south-west to Lhasa (see Map 10.1), changed the volume and the speed of these exchanges. Such trade opportunities shade over to seasonal and temporary migration, both for Tibetan peoples and for the semi-skilled Han people from Sichuan and other regions, looking for economic opportunities outside rural primary agriculture.

In modern China there are two official types of residence: either a permit for an urban area, the *Hukou*, or rural registration. Urban residents are subsidized at place of registration in terms of cheap grain, education, health and housing. Though subsidized grain is generally available in central Tibet, there otherwise are no such welfare facilities in rural areas. Until the reform era there was effective control by the local state apparatus over open mobility and residence. As has been noted before, this operated through the local registration of all people, through police supervision, and through the control of the distribution of food and clothing mainly through rationing at official work and neighbourhood locations.[69]

The general pressure in the People's Republic has been for rural–urban migration to try to take advantage of this material subsidy. The Civil Affairs Bureau and the Public Security Bureau are jointly responsible for rules and regulations on urban residence; current policy is to limit rural–urban migration and against the reclassification of rural areas as urban. The reason is that to increase the numbers categorized as urban would be to raise the numbers with a right to subsidized services from the state in towns, and thus increase the burden on the state. Reform policy stresses a greater role for the private market, not for the state. The number of urban migrants permitted each year has to be within the guideline fixed by these bureaux at a national level, which is usually 1.5 persons per 1000 already resident. For the TAR, the number is two per 1000 which implies that Lhasa City can add around 80 new registered residents per year with permits; Xining City, which is far larger, has an official quota of 200–300 people per year.

There may be those who would like to acquire urban residence registration in Tibet, but such registration would rarely be seen as an end in itself, more usually as a step in obtaining such residence in their original home areas. Priority in the allocation of these residence permits in any case is more likely to be given to senior war veterans, retired party cadres and the like, rather than to people from rural Sichuan who are wishing to trade in Tibet.

Reforms that give individual rights such as in choice of residence clearly work against limitations on population movement; after economic reform the market itself became an important force in population movements. Between the first implementation of reform in 1978, and May 1989, labour itself began to be treated in the capitalist manner as a commodity; that is, as a factor of production in economic development. Here labour, like any other factor of production, moves in response to market forces rather than directly reflecting state planning. The new ability to buy and sell goods and services through cash alone allows trades and other businesspeople to buy tickets for travel, rent accommodation, eat in restaurants and pay for these services directly, without the need for an accompanying written authorization from local work units. There is no open access to the subsidized accommodation, food, health and education services enjoyed by those with an official urban residence permit, but there is a new freedom of movement engendered by the market.

These reforms have had an impact in central Tibet. Before 1984 there were few places in public outside Lhasa where cooked food could be purchased with cash alone; even in 1985, people with permits tended to collect cooked food from county-level kitchens at which coupons rather than cash were the principal currency. Without written authority it was extremely difficult to obtain such coupons. For example, in 1981 an official trip from Lhasa westwards to Nepal would require prior arrangements with each of the counties to be crossed to obtain food, lodging or petrol. The existence of small restaurants and lodging houses, themselves a result of economic liberalization, together with open permission for temporary movement, have created the possibilities for private travel.[70]

However, there was no large-scale open movement of entrepreneurs into central Tibet, who preferred to move to cities on the eastern seaboard rather than to what was seen as a remote and backward area. Those who came are mainly from adjacent provinces. Few people from lowland China now would wish for rural residency rights in central Tibet; this is historically a different situation from that of the Sino-Tibetan borderlands themselves, where there is likely to have been a progressive encroachment on upland farming since the eighteenth century.

General growth in transport and urban construction has followed economic reform, and recent population movements of Han and Hui Muslim peoples into the plateau area are tied to the provision of these services in the non-formal sector. As described above, new migrants are almost exclusively in the urban areas, county headquarters and rural townships along the roads. They work in construction, in petty trade, and in semi-skilled service industries such as bicycle repair, hairdressing, restaurants and hotels.[71] There is also market-gardening of vegetables under polythene greenhouses in the hinterland of Lhasa, developed largely by Han Chinese, not by Tibetans. Han traders now rent land from Tibetans, paying Tibetans above the value of the grain quota the Tibetans are obliged to provide the state for the land; they then grow vegetables for market sale. It is only now, under economic reform and the household contract system, that Tibetans may dispose of their usufruct and rent their land to outsiders. In central Tibetan towns such secondary and service industries have a high profile, and in Lhasa the figures for registered Han residents is some 28.9 per cent of the town population; when supplemented by temporary visitors and non-civilians, this possibly would rise to between 50 and 60 per cent.

The semi-skilled Chinese workers in Tibet may not have so different a motivation to that of migrant labour elsewhere in Asia. Some have little economic opportunity in their home areas, carry out both routine work in which the indigenous population has little interest, and provide new services which are not present in the area. They are temporary residents without urban registration who may stay in Tibet on a temporary licence or permit. This is obtained from the Public Security Bureau of the county or municipality where they reside. Such a permit is quite easy to obtain, is given for a year at a time, and should be renewed after six months.[72] Some itinerant traders, like temporary visitors and other travellers, may not even have this document.

Newly-registered residents from the outside also include professionals such as doctors and engineers who have been transferred from lowland China. Their attitudes are not singularly different from the Westerners who work under contract as doctors or engineers in the Third World. They require higher salaries and better conditions than local residents to be induced to come to the high altitude plateau, in many cases resent having to work away from home, and do not regard their presence as permanent. Their term might be eight years; but if residence has been transferred to Tibet transferral back is by no means automatic.

Since state agencies have trouble in recruiting qualified professionals to maintain technical services in western areas, once recruited it becomes difficult for such people to obtain permission to leave. Employment has to be found in lowland China and arrangements have to be made with the Public Security Bureau and Personnel Office to have the residence permit and work documents transferred. A fee may be payable to the new urban authority where one will reside, and work units which might wish to offer employment may not be able to afford the fee, and so tend to appoint those who already have a residence permit.[73] Without the appropriate urban registration there would be no entitlement to the range of subsidized services that make it possible to live on a public employment salary.

Hence the pressure is not from professionals to move into Tibet but quite the reverse, to try to maintain residence in the home town in inner China. In 1991, the shift was to interim arrangements in which a move to Tibet would not commit a professional to a change of urban residence permit. Though there can be changes in *de facto* residence, registered residence could still be at a previous place of birth, education or residence. Families also may be split, especially between generations, and people try to ensure that their children are educated wherever their other family members are. No longer as in the early 1960s, they do not risk moving the entire family to Tibet. Hence a child's residence permit often will be at a home in lowland China, such as that of the grandparents. Extended terms of leave, which accompany professional postings in Tibet, allow for visits and for time to look for a work unit and transfer to lowland China.

ECONOMIC POLICY AND POPULATION MOVEMENTS

There are family groups of migrants in Tibet today, but it is more usual to see single men, or groups of men, isolated from their wives or families, or husbands and wives without their growing children. Whether professional or in service industry, many think of their presence as temporary. If these modern temporary population movements continue to be treated as such by the state, then it is a different order of phenomena to the earlier population movements into Tibet and Qinghai (and Xinjiang) which were often from direct food pressure, or were otherwise involuntary.[74]

The perceived high cost of new state settlement in urban areas can only be justified in terms of secondary economic growth. Yet though few urban residence permits are issued, these may be less relevant under continued market reform than at an earlier period of centrally planned allocation. Individual choice and indirect state motivation through economic incentives are now more important than direct state prescription alone, and are important features to movement into the Tibet Autonomous Region since the mid-1980s.

Besides the growth in urban service industries, there is another policy dimension to economic growth in the region, namely through increased trade across adjacent borders. Tibetans in border regions can pass through to trade in India and Nepal, as can local Kirghiz in Xinjiang to Kirghizia, with little restriction. Whatever the strategic reason for the establishment of the original settlements, and the current need for secure control of roads, on a day-to-day basis borders are opening to the passage of people and goods.[75]

The state has the development of a commodity economy as a priority, and since late 1992 has offered financial incentives for entrepreneurs, with tax relief, to set up secondary industry through classifying central Tibet as one of the 70 or 80 special economic zones in China. In 1992 there was again a policy debate within China on the western regions, between those who advocate training of local people as the key stage in economic development, and those who see the first step as bringing in outsiders from more developed areas to promote rapid economic change. If Tibetans cannot or will not take up this role, then Han and other peoples will be encouraged to move into the region.

Local people may well maintain their dominance in local border trade, but Han may well acquire a more dominant position in service industries which will be more dependent on exchanges with lowland China. The significance of these urban service industries is not the absolute numbers of non-Tibetan people *per se*, but the fact that this is the leading edge of the modernizing economy, and will become more important as the region modernizes. The numbers and relative position of Tibetan to Han in this sector is seen as important by Tibetans, and is a focus for expressions of civil discontent. Whatever their individual motivation, people in these industries eventually may become resident, either through new legislation or simply through economic changes that make official residence permits less relevant, and staying in a less-developed region more attractive. Temporary migration might become a *de facto* population movement.

In general, there tends to be a ratchet-like effect to population movements. It is possible that temporary residents in Qinghai and the TAR may, as in the past in Xinjiang, acquire more permanent residency rights, and that the market, rather than official residence certificates, will be the critical factor in defining such possibilities in the future. The final issue may be less the immediate wishes of migrants, than market forces and whether or not the state regards such population movements as temporary and reversible.

CONCLUSIONS

In the Tibet Autonomous Region and Qinghai there are settlements of Han people in the far-western strategic areas, and in the Tibet Autonomous Region also in the south-east, that date from the late 1950s and 1960s. Parallels can be drawn between the pattern of human change in modern Tibet, and that within the Chinese Empire at earlier periods, and there appears to be an overall continuity in political and administrative policy of the lowland state from traditional to modern times.

From at least the seventeenth century onwards, increasing population and the availability of New World crops encouraged Han migration into minority areas of western China, and began to shift the frontier of agriculture to hill areas westwards to the edge of the Tibetan plateau. Population movement was a means for civil control, a way to increase agricultural production, and allowed the perpetuation of the state. At that time the topography of the upland area of the plateau, and the associated problem of military control, still set a limit to agriculture and Han settlement.

The modern agricultural settlements and new towns of Qinghai and Tibet at first had a strategic rather than solely economic significance: they interposed a Han buffer between Tibetan and Mongolian peoples in Qinghai, and between minority peoples of Qinghai and those in Xinjiang. Civilian colonies following on from military garrisons are a well-known historical fact of imperial expansion in the region. There are regional precedents for political containment through control of external communications and border areas, with the British Empire in the far north of India. A pre-1947 political map of South Asia illustrates interposed strips of directly-administered British territory between the various Himalayan mountain states of Afghanistan, Kashmir, Nepal, Sikkim and Bhutan. These hill states, along with Tibet, may have been seen as potentially profitable locations for trade, but the primary political strategy was to regard the higher and more remote of these strips as buffer areas. Trade into such remote areas was not linked to a direct policy of local economic development or political integration, and there was no policy for population movement.[76] The agricultural and prison settlements to the far-east of the Stalinist Soviet Empire may well be more directly analogous, at least for Qinghai, which itself contains such prisons and forced labour camps.

The settlement of agricultural peoples from the Qing period onwards was accompanied by a general change in relative dominance within the civil order in the region from a transhumant pastoral 'tribal' state to a more sedentary, bureaucratic, state, based on agriculture. This reflects a more general tendency for the replacement of a nomadic 'horde' by the agricultural city-state.[77] The difference in the modern period is that technology has allowed another movement further upwards and westwards, with attendant changes in economic form and population distribution. One cumulative effect of this movement is that in the modern period pastoral peoples come to occupy marginal lands, not because traditionally they were located on such land or were socially marginal peoples, but because the extension of traditional states based on grain production and control of agricultural territory beyond the city walls has marginalized them.[78]

In Qinghai there have been more recent internal redistributions of population that have allowed Han people to move further westwards; in both regions migrants, whether temporary or otherwise, have also settled along roads and in large urban areas. Though Han are the clear majority in Qinghai, they are heavily localized in the lowland areas and Tibetans still clearly predominate in the south and south-west. To date, there has been no overall demographic supplanting of Tibetan by Han in the TAR, which is notable as the only autonomous region where the demographically predominant nationality is a 'minority' population.

China's policy between 1978 and 1992 has not encouraged migration and settlement in western areas in and of itself. It appears to have had two main objectives: one was the control and integration of local civil and community institutions under the state through the provincial and central government; the other has been economic growth within this structure through a mixed economy. For the state, security followed by economic growth are still the dominant issues, and though neither of these are necessarily linked to a policy for population movement, overall they may well encourage such movement indirectly. Economic growth through reform was first oriented towards more local administrative autonomy, and non-Han nationalities could be favoured here; but such a particular local regime appears to be politically tenable in the medium to long term only as long as these local regional governments follow centrally defined economic policy objectives.

After 1989, economic reform emphasized growth rather than diversity, and in Tibet, there was an ideological shift that emphasized conventional growth, and social change in line with lowland China, rather than growth according to local conditions.[79] In the Eighth Five-year Plan for the region,[80] growth was sought not just in primary agricultural production, but also in secondary industry and trade in central areas, where incentives were given for the investment of capital. The present policy is to encourage rapid growth, including trade with India and lowland China. In the short term, this policy of rapid change is likely to encourage skilled professionals and entrepreneurial migrants from lowland China to move into central Tibet, rather than to wait for Tibetans to be trained.

Overall, the key factor leading to population movement in western China at this time appears to be economic change. Indirectly, a policy of technological and economic change has promoted Han population movements into central and urban areas where they now predominate. Such migration appears not to be due to any particular Chinese policy for Tibet, or reflect a Han nationalism *per se*, but more a result of economic modernization. There are also likely to have been population movements into agricultural areas on the Tibet borderlands in Kham/Sichuan themselves, which have not been covered here, and which may well have a Han population similar to the level in eastern Qinghai.[81] Whether or not the existing Han farming population here is in a better position than the Tibetan to take advantage of market opportunities remains to be seen.

NOTES AND REFERENCES

1. 1990 population figures for ethnic minorities in China are 91.2 million. *Summary of World Broadcasts*, 27 March 1992, Far East, 1340, Section 4, Article 1, from *Lio Wang*, overseas edition in Chinese (Hong Kong), 2 March 1992.
2. Geographers refer to the Aihui-Tengchong line, an imaginary line drawn from Aihui in Heilongjiang province in the north-east of China, to Tengchong in Yunnan province in the south-west: half the land area, with 60 per cent of China's high mountains but only some six per cent of the population lies to the west of this line.
3. Figures on the total number of the Han population will vary according to whether the calculation includes travellers, temporary residents, and forces personnel, and level of mobilization of the last.
4. Field observations have been made for periods of over one month per year over a three-year period in central Tibet by a team of two to four in 1990, 1991 and 1992, as well as in 1986. They include Nyalam (also Nyanam Tt. *gnya'-lam*), Tingri (also Dingri (Tt. *deng-ri*), Sagya (also Sakya Tt. *sa-skya*), Shigatse (Tt. *gzhis-ka-rtse*), Bainang or Pa-nam (Tt. *Pa-nam*), Gyantse (Tt. *rgyal-rtse*), Rinbu (Tt. *rin-spungs*), Quxu or Chusu (Tt. *Chu-shur*), Lhasa, Maizhokunggar, (Tt. *mal-gro gung-dkar*), Lhünju (Tt. *lhun-grub*), Damxung

(Tt. *'dam-gzhung*), and DolungDechen (Tt. *stong-lung bde-chen*). (See note 9 below, on transcription.)

5. In normal travel, for Westerner or Chinese, vehicle stops will be along the main roads, at township and county headquarters. Such trunk-road stops rarely give a typical cross-sample of social life.
6. China Statistics Publishing House, *The Fourth Population Census of the Tibet Autonomous Region, Compilation of Survey Materials*, (Tianjin: Tianjin Municipality Beifang Publishing and Printing Company, 1990).
7. G.E. Clarke, 'Research Design in the Use of China's Census and Survey Data for Rural Areas and Households', in E. Vermeer (ed.), *From Peasant to Entrepreneur: Growth and Change in Rural China*, (Wageningen: Pudoc, 1982), pp. 217–40.
8. The following categories are officially included as resident in the 1990 records: those living in the TAR with a permanent residence permit; those present in the TAR for more than one year with a permanent residence permit elsewhere; those present in the TAR for less than one year, but who are away from the area where they have permanent residence permit for more than a year; those in the TAR for whom a permanent residence permit is pending consideration; those originally in the TAR as permanent residents but now working or studying abroad.
9. The abbreviation Tt. refers to the standard Tibetan transcription; see T.V. Wylie, 'A Standard System of Tibetan Transcription', *Harvard Journal of Asiatic Studies*, 22, (1959) 261–7.
10. Boundaries are as in Chengrui (ed.), *The Population Atlas of China*, Population Census Office of the State Council of the People's Republic of China and the Institute of Geography of the Chinese Academy of Sciences, (Hong Kong: Oxford University Press, 1987).
11. Ngari (Tt. *mnga'-ris skor-gsum*).
12. The 1982 census figures record that of the 3 864 388 people of Tibetan nationality and culture within the polity of China, 2 076 733 live immediately to the north and east of the TAR; this is in western Sichuan (922 024); Qinghai (754 254); Gansu (304 540); and Yunnan (95 915). *Annual Statistical Yearbook, Peoples' Republic of China* (Hong Kong: Oxford University Press, 1985).
13. Kham (Tt. *khams*), Amdo (Tt. *a-mdo*).
14. Tibet occupied the Chinese Capital in AD 763 and placed an emperor on the throne of Changan for a period of fifteen days. See L.I. Cuguevskii, 'Touen-Houang du vii au x siécle', *Nouvelle Contribution aux Étude de Touen-Houang*, Centre de Recherche d'Histoire et de Philologie de la iv section de l'École Pratique des Hautes Études II, Hautes Études Orientales, 17 (1981) 1–56.
15. D. Morgan, *The Mongols*, (Oxford: Blackwell, 1986).
16. The subsequent invasion of Southern Tibet by Nepal was in 1850. For local civil relations between south Tibet and north Nepal in this period see G.E. Clarke, 'The Great and Little Traditions in the Study of Yolmo, Nepal', in E. Steinkellner and H. Tauscher, (eds), *Contributions on Tibetan Language, History & Culture*, The Proceedings of the Csoma de Körös Symposium held at Velm-Vienna, Austria, Wiener Studien zur Tibetologie und Buddhismuskunde, 10 (1981) 21–37.

17. Chen Guangguo and Wang Haoxun, 'The Relationship between Mongolians and Tibetans during the Periods of the Ming and Chinese Dynasties and the Policy of the Central Chinese Government towards Mongolia and Tibet', *China Tibetology*, 1 (1989) 89–101.
18. Rockhill records in 1894 that the town population at Kumbun had increased slightly by traders settling from Sichuan and Shaanxi; W.W. Rockhill, *Diary of a Journey through Mongolia and Tibet in 1891 and 1892* (Washington, DC: Smithsonian, 1896), p. 70.
19. E. Teichman, *Travels of a Consular Officer in East Tibet*, (Cambridge University Press, 1922), and H.J. Wiens, *Han Chinese Expansion in South China*, (Hamden, Conn.: Archon Books/ Shoe String Press, 1967), p. 192.
20. J.D. Durand, 'The Population Statistics of China', *Population Studies*, XIII (1960) 209–57.
21. Overall, though population trebled, land availability only doubled, which explains the pressure for intensification of agriculture. J. Banister, *China's Changing Population*, (Stanford, Calif.: Stanford University Press, 1987).
22. W.W. Rockhill, cited in Ho Ping-ti, *Studies on the Population of China*, (Cambridge, Mass.: Harvard University Press, 1959).
23. Maize and sweet potato in all probability came from India through Burma to Yunnan in the late sixteenth century; maize is recorded in Yunnan in 1574.
24. Ho Ping-Ti, *Studies on the Population of China*, pp. 137–48, 229.
25. Golog (Tt. *'go-log*, *'gu-log* or *mgo-log*; pinyin, *Guoluo*). See J.F. Rock, *The Amnye Ma-chen Range and Adjacent Regions*, (Rome: Istituto italiano per il medio ed estremo oriente, Serie orientale, 1956).
26. L. Gelek, 'The Organisation of the Nomads of Wasu-Sethar', (Oxford: unpublished manuscript, 1992).
27. This unification was carried out by the local leader Gonpo Namgyal between 1835 and 1865, J. Norbu, *Warriors of Tibet: The Story of Aten and the Khampas' Fight for the Freedom of their Country*, (London: Wisdom Press, 1979).
28. One key issue in current disputes on Tibet's sovereignty is to what degree, after the Han Chinese army took Lhasa in 1720, the Tibet government continued to exercise authority in its own right rather than as a 'suzerain' of China. For an account of border formalities at this period see A. Heller, 'Mi Dbang's 1740 Decree to Batang', *Acta Orientalia Academiae Scientarium Hung Budapest*, XLIII, 2–3, (1989), 375–90; for the later period see M.C. Goldstein, 'The Dragon and the Snow Lion: the Tibet Question in the Twentieth Century', in A.J. Kane (ed.), *China Briefing*, (Boulder, Colo.: Westview Press, 1990).
29. In principle, after the founding of the Chinese Nationalist state, Kham came under the authority of the Army of General Lie Wu Hen, and Amdo came under the Muslim General Ma Bufung based in Xining to the north. Ma Bufung finally was expelled from Qinghai in 1938, and not by the People's Liberation Army, but by two tribes of Golog, the Kang-gan and Kan-sai; see Foreign Affairs Office of Qinghai, *Qinghai Province*, (China: People's Government of Qinghai, 1979), p. 8.
30. See G.E. Clarke, 'China's Reforms of Tibet, and their Effects on Pastoralism', *Discussion Paper 237*, (Brighton: University of Sussex, Institute of Development Studies, 1987).

31. The total population of Qinghai is recorded as 4 215 890; all figures presented for Qinghai refer to the 1989 population records, unless otherwise stated. The north-eastern area as defined here is Haidong Prefecture, Xining Municipality and Daotong Municipal Area: it contains 67.6 per cent of the population, that is some 2 848 753 people (Table 10.1).
32. The modal range for Tibetan population by prefecture is between 100 000 and 200 000.
33. Tu form 3.4 per cent, Salar 1.6 per cent, and Mongolian 1.5 per cent, of the population of Qinghai.
34. Ninety-three per cent of the total Salar population of 68 509 are in Hualong and Xunhua counties; 61 per cent of the total Tu population of 141 770 are in Minhe and Huzhu counties; these counties are all in Haidong prefecture. The material culture of the Tu and Salar is related to that of the Tibetan and Mongolian peoples of the area; the Hui Muslims are a more 'open' group into which Han can marry and be included, and their language and material culture, at least today, are related to that of the Han majority.
35. The arithmetic mean for Tibetans in each of the county-level units (of which there were some 43 in 1989 including Xining and Daotong municipalities) is approximately 19 600; a preliminary banding gives five counties with less than 1000 Tibetans, seven with between 1001 and 10 000, 14 with between 10 001 and 20 000, six with between 20 001 and 30 000, six with between 30 001 and 40 000, and five with over 40 000.
36. For example, in 1989, of the 12 large state farms to the south of the main highway close to Lake Qinghai, half were enterprises run directly by the army itself.
37. G.E. Clarke, 'Aspects of the Social Organisation of Tibetan Pastoral Communities', in I. Shoren and Z. Yamaguchi (eds), *Proceedings of the 5th Seminar of the International Association for Tibetan Studies, 1989* (Japan, Narita: Naritasan Shinshoji, 1992), pp. 393–412.
38. Clarke, 'China's Reforms of Tibet'.
39. Golmud (Pinyin, Geermu); Caidam (Tt. *tsha'i hdam*), also Da Caidam, Dachaidam, Daqaidam, in some transcriptions.
40. Data are not yet available on Lenghu County for 1989 for the overall increase in population by nationality.
41. The population of Xining City, which is larger than the urban area, is given as 550 316.
42. For the 1930s see E. Maillart, *La Vie immediate*, (Lausanne: Payot, 1991), p. 112.
43. Licences were required to fish in 1982/3; and overall quotas imposed from 1985 effectively banned private fishing other than possibly by local communities. Clarke, 'Notes on the Socio-Economic Impact of Qinghai Fisheries', *Report to UNDP/FAO*, (Beijing: 1989).
44. Approximately one-third of the 500 newly-settled households above the Yellow River Gorge in Hainan in 1989 were Tibetan, this group having their own irrigated step on the hillside.
45. Nying-tri (Tt. *nying-khri*) also Nying-che, Nyingti, Ninji, or Ningche following the traditional Tibetan name of the major town. Also commonly Linji (Pinyin, *Linzhi*), Ling-ji, or Lin-chih. Distinct from the county of Lhünju (Pinyin, *Linzhou*) in Lhasa municipality, and the county of Lhünze (Pinyin,. *Longzi*) in Lhoka (Shannan) prefecture.

46. The registered population figures are: Nying-tri County – 8634 Han, 20 678 Tibetan, 764 Mönba, and 74 Hui Moslem; Po Mi County (also known as Po Bho, Tt. *po-bo* ?) – 2460 Han and 22 273 Tibetans; Ming Ling (Tt. *smin-gling*, Pinyin, *Milin*) County – 1574 Han, 12 507 Tibetans, and 930 Löba.
47. Radio Lhasa, 15 July 1983.
48. D. Choedon, *Life in the Red Flag People's Commune*, (India, Gangchen Kyishong Information Office, Dharamsalaa: 1978).
49. Metog also Medog (Pinyin, *Muotuo*); Zayu (Tt. *rdza-yul*, Pinyin, *Chayu*). 'Cona' county to the west of Metog and bordering eastern Bhutan in parts covers the same territory as the Miri and Dafla Hills of Arunachal Pradesh.
50. See A. Lamb, *The McMahon Line: A Study in the Relations between India, China and Tibet, 1904–1914*, (London: Routledge, 1966).
51. N. Maxwell, *India's China War*, (London: Jonathan Cape, 1970).
52 It may well be because of this military stand-off that Metog still has no full all-weather road access from China or India, and retains its timber and a full range of Himalayan bio-diversity.
53. The recorded population figures are: Za-yul county (total 22 801) 705 Han, 20 486 Tibetan, and 1610 'other'; Metog County (total 8714), 79 Han, 1435 Tibetan and 7200 Mönba or Löba.
54. Mönba refers generally to people from wooded areas south of the Himalaya and is used as a proper name in this area; Löba usually is used to refer to people from Bhutan. See C. von Fürer-Haimendorf, *Highlanders of Arunachal Pradesh: Anthropological Research in North-East India*, (Delhi: Vikas, 1978), and his earlier works for an ethnographic account of such peoples.
55. The Tibetan state itself expanded in this way into these tribal areas to the south-east of the plateau, such as those of the Na-khi, at various times from the 9th century onwards.
56. Senge Kebab is Tibetan ('lion's mouth', referring to the source of the Indus); Gar (Pinyin, *Geer*, Tt. *sgar*); the old capital was in Purang, which has only a five per cent registered Han population.
57. Yuan Qingli, 'Population Changes in the Xinjiang Uygur Autonomous Region (1949–1984)' *Central Asian Survey*, 9, 1, (1990) 49–73.
58. The total number of official residents listed in Ngari (A-li) prefecture is 61 639, with 1153 Han and 8212 Tibetan in Gar county.
59. Interpreting the data in this prefecture is difficult: whereas the figures for the number of Han and Hui Muslims given for 1992 are the same as those for 1990, the figure given for Tibetans (542 209 and 559 992) and others, largely Sherpa (Xiaerba) at (1866 and 1887), register annual percentage increases of 1.64 per cent and 0.51 per cent over these two years.
60. In a village in the pastoral south of south-east Shigatse prefecture near the Sikkim border, where no Westerner would have been likely to have passed through since the first decade of this century, as I stepped down from a jeep I heard one Tibetan ask another whether or not I was a Chinese.
61. There were larger Chinese military camps before 1971, and the current county Headquarters of Pa-nam is located at just such a deserted military base along the main road.
62. See E.G. Mihaly, *Foreign Aid & Politics in Nepal*, (Oxford: Oxford University Press, 1965).

63. On human depredations in Tibet see, for example, Legal Inquiry Committee of Tibet of the International Commission of Jurists, *Tibet and the Chinese People's Republic*, (India: Sterling, 1966).
64. An analysis of age-distributions of post-1982 data more detailed than can be carried out at the present time would be required, looking for an ageing cohort smaller than that suggested in a normal Gaussian distribution, and possibly a skewed gender distribution with a lower number of males within that ageing cohort. Earlier Tibetan population estimates themselves vary at their widest by an order of magnitude, and otherwise between two and six million. Comparison of these with subsequent Chinese statistics presents obvious methodological problems. Generally, little pre-1982 data can be taken at face value in China; this is especially the case for remote, poorly-administered, areas in times of conflict. See Clarke, 'Research Design in the Use of China's Census and Survey Data for Rural Areas and Households', in E. Vermeer (ed.) *From Peasant to Entrepreneur: Growth and Change in Rural China*, (The Hague: PUDOC/ Centre for Agricultural Publishing and Documentation, 1991) 217–40.
65. J. Dreyer, 'China's Minority Nationalities in the Cultural Revolution', *China Quarterly*, 1968, no. 35, July–Sept, p. 108.
66, Malnutrition is indicated by reports from refugees, from reports of earlier exiles subsequently returned to Tibet, from internal reports, and by the various nutritional deficiencies recognizable today among the adult population. For example, see T. Tersey, 'One Month in Tibet', *The Tibet Society Newsletter*, New Series, 8, (Bloomington, Ind.: 1979), pp. 1–36.
67. Clarke, 'China's Reforms of Tibet'.
68. Official statistics indicate that the number of Han residents dropped from 105 849 to 76 300 between 1982 and 1984.
69. For example, see J. Spence, *The Search for Modern China*, (London: Norton, 1990), p. 688.
70. For example, when I first visited Damxung county in 1985, there was only one central county kitchen; but by 1991 there were 12 small restaurants and various other shops, services and lodging houses, and the central county kitchen had closed down.
71. Some reports suggest that in 1980 Tibet's 200 registered business enterprises and factories were staffed mainly by Han nationals; in Qinghai in 1989, a Tibetan carpet factory south of Lake Qinghai was staffed by Han weavers from the eastern seaboard.
72. In 1989 the initial charge was ten yuan, plus a fee of six yuan for renewal (that is, under 5 per cent of a monthly state salary).
73. In 1991, for Shanghai, the figure that had to be paid to the municipality was around Y40 000; for Chengdu it was Y6000.
74. Xinjiang provides another example of Han population movement in modern times. The area was brought under control of the Qing after 1884 at which time there were very few if any Han peoples resident. The armies which helped to establish control for the PRC post 1949 stayed on as settlers, and other settlers were encouraged to the area from lowland China. According to the 1990 figures, Han settlers (including state-farm workers, ex-prisoners and military), formed around 43.5 per cent and Uygur 55.0 per cent of the popu-

lation of Xinjiang. Yuan, 'Population Changes in the Xinjiang Uygur Autonomous Region.'

75. A border trade agreement was reached in 1992 between India and China and led to the reopening of the old route from Gyantse south to Sikkim, and in the west the pilgrimage route from Mount Kailash.
76. Clarke, 'Who were the Dards? A Review of the Ethnographic Literature of the North-Western Himalaya', *Kailash, Journal of Himalayan Studies*, 5, 4 (1977), 323–56.
77. D. Morgan, *The Mongols*, (Oxford: Blackwell, 1986).
78. See P. Rigby, *Persistent Pastoralists: Nomadic Societies in Transition* (London: Zed, 1985), p. 96; also P. Bonte, 'Marxist Theory and Pastoralist Societies' in J.S. Kahn and J.R. Lobera (eds), *The Anthropology of Pre-Capitalist Societies*, (London: Macmillan, 1981), pp. 22–56, for an attempt at a social analysis of the relations of pastoral to sedentary societies based largely on East African material.
79. Tibet Youth Association for Theory, 'Reflections on the Current State of Theoretical Policy on Tibet', *Tibet Daily*, 7 August 1989.
80 Government of the Tibet Autonomous Region, *The Tibet Autonomous Region Five and Ten Year Development Plans (Economic & Social)*, (Lhasa, Tibet, 1992).
81. Clearly, many of the same factors currently operate in Kham; unfortunately, data at hand do not allow us to determine the boundaries and scale, or to distinguish between pre-1951 and post-1951 movements of Chinese people.

Index